Ground Water

GROUND WATER

Editor

P. SANKARA PITCHAIAH

Department of Geology
Nagajuna University
Nagarjuna Nagar - 522 510, India

SCIENTIFIC
PUBLISHERS

Published by:
Scientific Publishers
5A, New Pali Road, P.O. Box 91
Jodhpur 342 001, India
E-mail: info@scientificpub.com
Website: www.scientificpubonline.com

Print : 2022

ISBN : 978-81-72330-71-2
eSBIN : 978-93-88449-74-8

Printed in India

CONTENTS

1

INTRODUCTION

Dr. P. SANKARA PITCHAIAH

Department of Geology, Nagarjuna University

One who solves the problem of water is worthy of two noble prizes, one for peace and one for science.

— John F Kennedy (1917-1963)

Water is next to air, required for the support of life. Being elan vital, it figures in all the phases of man's activity. The world's total water resources are estimated at 1.37×10^8 million ha-m and their distribution is as follows;

Saline water in Oceans	-	97.2%
Land areas	-	2.8%
Ice caps and glaciers	-	2.14%
Groundwater	-	0.61%
Soil moisture	-	0.005%
Fresh water lakes	-	0.009%
Rivers	-	0.0001%
Saline lakes	-	0.008%
Atmosphere	-	0.001%

India, having a geographical area of 329 m-ha and supporting a population of 845 million (1991) is a land of climatic contrasts. The season and weather conditions range from extreme cold to dry spell to extreme humidity and from drought conditions to torrential rains and floods.

GROUND WATER

Editor: P. Sankara Pitchaiah, © 1994 Scientific Publishers, Jodhpur

The average annual rainfall of India is 117 cm. Dr. K.L. Rao estimated the contribution of rainfall to groundwater as 37 M. ha-m. According to the National Commission on Agriculture total precipitation is 400 M. ha-m. Out of this 115 hac-m, M. is the surface water and 50 M. ha-m. is the groundwater. 20. M. of surface water has been shown as flow from out side India and thus the total annual basic water resources of the country has been estimated as 185 M. ha. m.

Hydrogeological Divisions

Central Ground Water Board has evolved following categories to describe the hydrogeological characteristics of various rock types occurring in the Country.

I.Porous Formation:

(a) Unconsolidated formations : The quarternary sediments - Recent alluvium, older alluvium and coastal alluvium form the important aquifers. These are composed of clays, silts, sands, gravels, pebbles, cobbles, boulders, ferruginous nodules, kankar, etc.

The piedmont zone of the Himalayas is skirted by artesian belt under free flowing conditions extending from Punjab to Assam. The hydrogeological environment and groundwater regime conditions in Indo-Ganga-Brahmaputra Basin indicate the existence of fresh groundwater reservoir about 600 m or more below land surface.

(b) Semi-consolidated formations : The semi-consolidated formations belong to Paleozoic, Mesozoic and Cenozoic group of rocks extending from Carboniferous to Mio-pliocene in age. These are chiefly composed of shales, sandstones, limestones, flysch and molasse beds. The terrestrial fresh water deposits belong to Gondwana system of the peninsular shields are also included under this category. Under favourable conditions these formations give rise to flowing conditions as in parts of Godavari valley, Vallar basin, Cambay basin and parts of West coast.

II Fissured Formation :

Consolidated formations : The consolidated formations occupy almost two thirds of the country. Most of the consolidated rocks, except vesicular volcanic rocks have negligible primary porosity. From the hydrogeological point of view, the fissured rocks are broadly classified into the following four types.

(i) Igneous and Metamorphic rocks excluding volcanic and carbonate rocks : The most common rock types are granite, gneisses, charnockite, khondalite, quartzite, schist, phyllite, slate, etc. Groundwater yield depends on rock types for ex: granite, khondalite and biotite gneisses are better sources than charnockite. Grade of metamorphism is also an important factor. These rocks rendered porous due to the secondary

openings by weathering and fracturing.

(ii) Volcanic rocks : The Deccan Trap lava flows are mostly horizontal and the groundwater occurrence is controlled by the contrastic water bearing properties of different flow units. The topography, nature and extent of weathering, jointing and fracture pattern, thickness and depth of occurrence of vasicular basalts are the important factors which play a major role in the occurrence and movement of groundwater in these rocks.

(iii) Consolidated sedimentary rocks excluding carbonate rocks : Consolidated sedimentary rocks occur in Cuddapah, Vindhyans and their equivalents. The formations consist of conglomerates, sandstones shales, slates and quartzites. Locally they contain phyllites and schists. The occurrence and movement of water in them is governed by bedding planes, cleavages, fractures, joints, contact zones and magnitude of weathering, topography and climate.

(iv) Carbonate rocks : Carbonate rocks include limestone, marble and dolomite. Potential karstified aquifers are found to occur in Rajasthan and Peninsular India. Springs are reported from the limestones of Himalaya region.

Groundwater Provinces of India

Groundwater provinces are the regions having the general similarity in the mode of occurrence of groundwater. Taylor (1959) grouped various stratigraphic units of India into eight groundwater provinces (Table-1).

Water Resource Development in India

For a controlled development, proper management of a resource and its conservation the first requirement is a knowledge of the total resource potential. Unfortunately there is a large gap at this stage in India (Ramesam, 1989). The first approximate estimate was made in the late forties and all the subsequent efforts were only slight improvements and adjustments. As a result, precise quantitative data on precipitable moisture content in the monsoon winds, actual evaporation losses in each river basin, variations in the glacial melt water from the Himalaya and many other issues are glaringly absent. Proper collection and preservation of resistivity survey data, quality analysis and well inventory data are lacking. Neverthe-less, progressive development in harnessing the water resources has being going on. The total investments in exploiting the water resources for not only irrigation but other purposes of water use are given in table-2.

Table - 1. Groundwater Provinces in India

Province	Age	Systems	Rocks	Area (sq.km)	Distribution	Groundwater occurrence	Wells	Yield
Pre-Cambrian Crystalline province	Pre-Cambrian to Archaean	Delhi system. Dharwar and Aravalli	quartzite, phyllite, schist, calc-gneiss, granite, gneiss, schist, amphibolite, quartzite, marble	1,000,000	Kanyakumari to Delhi	Weathered and fractured zones	Dug wells, bore wells	Generally below 25 m^3/h
Pre-Cambrian Sedimentary province	Pre-Cambrian	Vindhyans and Kurnool systems and equivalents, Raipur and Indravati series	limestone, shale, quartzite, sandstone conglomerate	210,000	Watersheds of chambal, Yamuna, Son and Narmada rivers of Rajasthan, Uttar Pradesh, Madhya Pradesh and Bihar	Karst terrains, Weathered zones bedding planes joint openings, solution passages	Dug wells, bore wells	5-200 m^3/d
		Cuddapah system, Kaladgi, pakhal and Bijawar series	Inter-bedded limestones, shale, quartzite, sandstone and conglomerate.		Greater part of Rayalaseema region in Andhra Pradesh and isolated areas in Karnatake.			
Gondwana Sedimentary province	Late Carboniferous to Early Cretaceous	Gondwana system Talchir, Damuda, Panchet, Mahadeva Rajamahal, Jabal pur, Umia series, Himmatnagar sand stones and Lathi series	Sandstone and shale, tillite		Godavari, Mahanadi, Narmada, Son and Damodar valleys of Central India, Western Rajasthan, Kutch, Saurashtra areas and scattered patches along the east coast.	Flowing wells, semi-consolidated rock aquifers	Dug wells, tube wells	100-400 m^3/d

Table Contd...

Province	Age	Systems	Rocks	Area (sq.km)	Distribution	Groundwater occurrence	Wells	Yield
Deccan Trap province	Upper Cretaceous to Lower Eocene	Deccan traps	Basalt flows and inter trappean beds, tuffs, breccia and ash	500,000	In parts of Maharashtra, Karnataka, Madhya Pradesh, Andhra Pradesh, Gujarat and Rajasthan	Weathered and jointed horizons, contact zones	Dug wells, bore wells	20-125 m^3/d
Cenozoic Sedimentary Province	Eocene to Pliocene	Palana and Nagaur sand-stones in Rajasthan; Cuddalore, Rajahmundry and Athgarh sandstones in the east coast, Warakalli and Dwaraka beds in the west coast	Sandstone and shale	–	Narrow coastal plains on the Malabar and Coramandal coasts, coastal fringes on Saurashtra and Kutch peninsulas, and extensive embayment in Gujarat and Rajasthan and a belt of strongly folded rocks in Eastern India	Confined and unconfined conditions	Dug wells, filter points	50-150 m^3/d
Cenozoic Fault Basins	Pliestocene to Recent	Narmada, Tapti river alluvium	Gravel, sand silt, clay	–	Narmada, the puma and Tapti valleys	Un-consoli dated sediments, under confined and un-confined conditions	Dug wells, filter points	25-150 m^3/d
Ganga, Brahmaputra Alluvial province	Pleistocene to Recent	Alluvium in the fore deep between the Himalayas and Peninsula	Boulder, gravel, sand, silt, clay	850,000	Vast plains of Ganges and Brahmaputra rivers	Aquifers with unconsolidated sediment	Dug wells, Dug-cum-bored wells, bore wells, cavity wells	30-300 m^3-h
Himalayan Highland Province	Archaean to Recent	Varied	granite, sandstone, slate, shale, limestone	–	Kashmir to Arunachal Pradesh	Weathered and fractured zones, valley portions	Dug wells, tube wells	100-200 m3/h7

Table 2. Plan-wise investments in exploitation of water resources

(Rupees in crores)

S. No.	Period	Urban Water supply	Rural Water supply	Irrigation	
				Major	Minor
1.	First Plan (1951-56)	—	33	380	66.00
2.	Second Plan (1956-61)	—	—	380	161.15
3.	Third Plan (1961-66)	89	67	581	443.29
4.	Annual Plan (1966-69)	39	21	434	560.74
5.	Fourth Plan (1969-74)	261	131	1237	1174.06
6.	Fifth Plan (1974-78)	539	381	2442	1411.24
7.	Annual Plan (1978-80)	—	—	2056	987.40
8.	Sixth Plan (1980-85)	1419	1635	7516	3239.55
9.	Seventh Plan (1985-90)	2555.10	4470.41	10986.06	6426.91
10.	Annual Plans (1990-92)	1634.69	2586.12	5697.69	—
11.	Eighth Plan (1992-97)	5647.28	10054.52	—	—

REFERENCES

Ramesam, V., (1989) Research and development in water resources. J. Geol. Soc. India, V. 34, pp. 345-350.

Taylor, G.C., (Jr) (1959) Groundwater provinces of India. Econ. Geol. V. 54, pp. 683-697.

2

HYDROCLIMATOLOGY

H. S. RAM MOHAN and K. SHADANANAN NAIR

Physical Oceanography & Meteorology Division,
School of Marine Sciences,
Cochin Univ. of Science & Technology,
Fine Arts Avenue, Cochin - 682 016.

Hydrology and meteorology share many areas of common concern such as the hydrological cycle, the thermal and water balances at the land surface and their interactions, and the different global chemical cycles. The science of hydrology deals with the occurrence and distribution of water on land, its precipitation from the atmosphere, its storage, run-off and evaporation, and its ultimate return to the atmosphere and ocean. Hydroclimatology is that part of hydrology concerned primarily with the study of atmospheric and land phases of the hydrologic cycle, with emphasis on the inter-relationships involved.

At any given time, the oceans contain about 97.2% of the total world water supply. The balance is fresh water of which 2.15% is in ice caps and glaciers, 0.625% in soil and ground water, 0.017% in surface water and only 0.001% in the atmosphere. The atmospheric water vapour is equivalent to a layer of water 2.5 cms high over the entire globe. Most atmospheric moisture comes from the sea through a complex exchange of energy and matter between the ocean and atmosphere.

The atmospheric phase

It is in the cycle of moisture between ocean and land that the hydrologic cycle appears as an integral feature of large atmospheric circulations. The entire process is an intricate combination of evaporation, transpiration, air mass movement, condensation, precipitation and run-off.

GROUND WATER
Editor: P. Sankara Pitchaiah, © 1994 Scientific Publishers, Jodhpur

Evaporation

The hydrologic cycle may be considered to begin with evaporation from the ocean and other water bodies and transpiration from vegetation. Evapotranspiration (or evaporation and transpiration) is the key element of hydroclimatology as it is the link between moisture and energy exchanges. Evaporation has also an important role in global heat balance, as it utilizes large amounts of latent heat which is later released during condensation. Evaporation depends on a number of environmental factors - differences in vapour pressure at the evaporating surface and air above, air and water temperatures, wind, atmospheric pressure, nature of evaporating surface (soil, vegetation, pressure, etc.) and even quality of water. Transpiration is influenced by all the above factors; in addition, physiological factors such as density and behavior of stomata, extent and character of protective covering, leaf structure and plant diseases also influence it.

Because evaporation and transpiration play such an important role in the global hydrologic cycle, a great deal of effort has been expended in attempts to measure them directly or estimate them indirectly either singly or together.

Direct measurement of evaporation or evapotranspiration from large water bodies or land surfaces is not possible at present. However, several indirect methods such as evaporation pans and lysimeters give acceptable results. Estimates can also be made by the energy budget method, the aerodynamic approaches or by the water budget method. Bowen (1926), McEwen (1930), Richardson (1931) and Cummings (1936) used formulae based on energy budget methods while the aerodynamic approach was employed amongst others by Sutton (1934), Sverdrup (1937) and Thornthwaite and Holzman (1942). A combination of aerodynamic and energy balance methods was suggested by Ferguson (1952), Penman (1956) and Slatyer & McIlroy (1961).

Thornthwaite (1948) introduced the concept of Potential Evapotranspiration (P.E.) as the maximum amount of water lost to the atmosphere from a large surface covered with vegetation and where there is no shortage of soil moisture at all times for full and uninhibited use. This unique definition makes P.E. dependent only on meteorological parameters such as radiation, temperature, wind and humidity and independent of nature of evaporating surface, vegetation or cultural practices.

On the other hand, Actual Evapotranspiration is the amount actually available for evaporation and transpiration and depends on P.E., precipitation and soil moisture content.

Many formulae are available to estimate P.E. of which the one by Penman (1956) has found world wide acceptance. Thornthwaite established a relation between mean monthly temperature and P.E. Later,

Thornthwaite and Mather (1956) suggested a book-keeping procedure of water budgeting which estimates A.E. among other parameters.

Precipitation

The atmosphere thus gains moisture from evapotranspiration. The amount of water vapour it can hold is a function of temperature : the warmer the air, the greater its water holding capacity. When its capacity is reached, it is said to be saturated. The precipitation process begins with the formation of clouds which are composed of water droplets or ice crystals or both formed by condensation or sublimation of water vapour on condensation nuclei. These cloud droplets grow in size and precipitate as rain, snow or hail.

The Land Phase

Precipitation is the most widely measured hydroclimatic element and its measurements are within acceptable limits of accuracy. The parameters of interest here are the rainfall amounts, intensity and the number of rainy days. In many hydroclimatic problems, the spatial and temporal distributions of precipitation are important. Intense rainfall, even of short durations, can cause soil erosion while prolonged heavy rainfall may cause floods. On the other hand, deficient rainfall could lead to water deficits and droughts.

A part of the precipitation falls directly on the oceans and the remainder on the land surface. A portion of water falling over the land is intercepted by vegetation or structures and may evaporate later. The other portion falling over the land unintercepted, infiltrates into the soil to form a thin layer of water, close to the surface, called soil moisture. This may either evaporate into the atmosphere or when the storage capacity of the soil (field capacity) is exceeded, may percolate downward and join groundwater. When the rate of precipitation is heavy and exceeds infiltration rate, surface runoff occurs to form rivers, lakes or streams, all of which finally join the sea. Groundwater, too, flows through subsurface channels to join the sea ultimately.

Precipitation in the form of snow is different as it may accumulate on the earth's surface for some time before it melts and runs off to streams or re-evaporates into the atmosphere.

Therefore, all the water evaporated into the atmosphere is returned to the oceans or other water bodies, thus completing the hydrologic cycle. The exact nature of this closed chain of events at any given region determines its hydroclimatology.

The water balance model of Thornthwaite and Mather (1955) is of great practical use in studying the hydroclimatology of any region by comparing precipitation (water supply) with potential evapotranspiration (water need) making allowance for soil moisture storage. For a given station, if precipitation always exceeds P.E., the evaporation is at the potential rate, the soil remains at field capacity and a water surplus occurs. When precipitation is less than P.E., the actual evaporation is equal to the sum of the precipitation amount and the moisture available from soil. As soil dries out, it becomes more difficult to utilize the soil moisture. When actual evapotranspiration is below P.E., a water deficit occurs.

In other words, water deficiency is the amount by which precipitation and soil moisture together fail to meet the demand of P.E. Water surplus, on the other hand, represents the excess of precipitation after meeting demands of P.E. and recharge of soil storage to field capacity. The surplus water appears as either surface runoff which joins lakes and rivers or as subsurface runoff which percolates to join groundwater depending on surface and subsurface physiographic features. Part of the surplus water is detained and the remaining part flows as run off : the amount of runoff depends on slope, soil type and vegetation in the area.

The water balance approach to the study of hydroclimatology in India was first attempted by Subrahmanyam (1956). The concept has now established itself in many fields like classification of climates (Subrahmanyam 1956), Subrahmanyam and Ram Mohan (1980), drought assessment (Subrahmanyam (1967) Subrahmanyam and Sastri, (1971,) Ram Mohan (1978,) Ram Mohan et al., (1984) estimation of yields from river basins (Subba Rao & Subrahmanyam (1961,) and Agricultural Planning (Ram Mohan & Subrahmanyam (1984)). Details of the model and its application have been given by Subrahmanyam (1982).

Hydroclimatological studies are a prerequisite in developmental planning of any area. The prosperity and growth of any region is related to the available water resources which in turn is related to the climate.

CASE STUDY WITH REFERENCE TO INDIA

Introduction

India is a land of vast diversity in culture, language and custume. Such a diversity is observed in its physiography, climate and water resources, too. The lofty Himalayas in the north, broad plains below them, the hill ranges to the north-east and the lower range running along the west coast of the peninsula all determine the climate of India to a great extent. The rainiest region in the world happens to be in the eastern portion of the country; ironically, in the same latitude belt, the desert dominates in the west. The annual range in mean temperature is only 4°C in some of the coastal areas, while it is as high as 50°C in some interior regions.

Rainfall distribution over India

The Indian agriculture and economy are largely dependent on the southwest monsoon rainfall. The southwest and northeast monsoons and the local weather systems produce an average annual rainfall of about 110 cms. Orography results in an annual rainfall of more than 400 cms in the north-eastern part of the country and in the region west of the Western Ghats (Fig. 1). The station, Cherrapunji, on the southern slopes of the Khasi-Jaintia Hills receives more than 1000 cms of rainfall in an year and Amboli (16°N) in the Western Ghats gets more than 740 cms. East of the Ghats, rainfall decreases to 60 cms and then slightly increases to about 100 cms towards the east coast. The region of lowest rainfall in the peninsula-less than 50 cms-is found around 18°N on the eastern side of the Ghats.

From the coast of West Bengal and the hills of Orissa, rainfall decreases inland, in a north-west direction. Over the western parts of the country, rainfall decreases to even less than 20 cms in an year. Rainfall in the Himalayas increases from around 200 cms in the west to more than 400 cms in the east.

The islands of Lakshadweep in the Arabian Sea get only 150 cms of rainfall whereas the Andaman and Nicobar islands in the Bay of Bengal get more than 300 cms.

Number of Rainy days

The pattern of rainy days is similar to that of rainfall (Fig. 2). Cherrapunji has the maximum number - 160 rainy days in any year. The eastern parts of the country and the west coast have 150 cms of rainy days, but in the interior, the days are considerably less. In the north-western parts, there are less than 20 days rainfall, (Rao, 1981). The average rainfall over the plains is 1.5 to 2 cms per rainy day (Pisharoty, 1986).

Seasonal Distribution of Rainfall

The southwest monsoon season (June-September) is the principal rainy season for the country and therefore, the seasonal distribution pattern closely resembles the annual (Fig. 3). Orography produces heavy rainfall (upto 250 cm) along the west coast, in the foothills of the Himalayas and in the eastern portion of the country. Rainfall is comparatively less in the interior peninsula, northern plains and in the north-western region.

The picture is different in the north-east monsoon period (October-December). The chief areas of rainfall in this season are confined to the southern tip of the peninsula, the east coast, Assam and parts of Kashmir. A north-west to south-east increase in rainfall is noticed in this season.

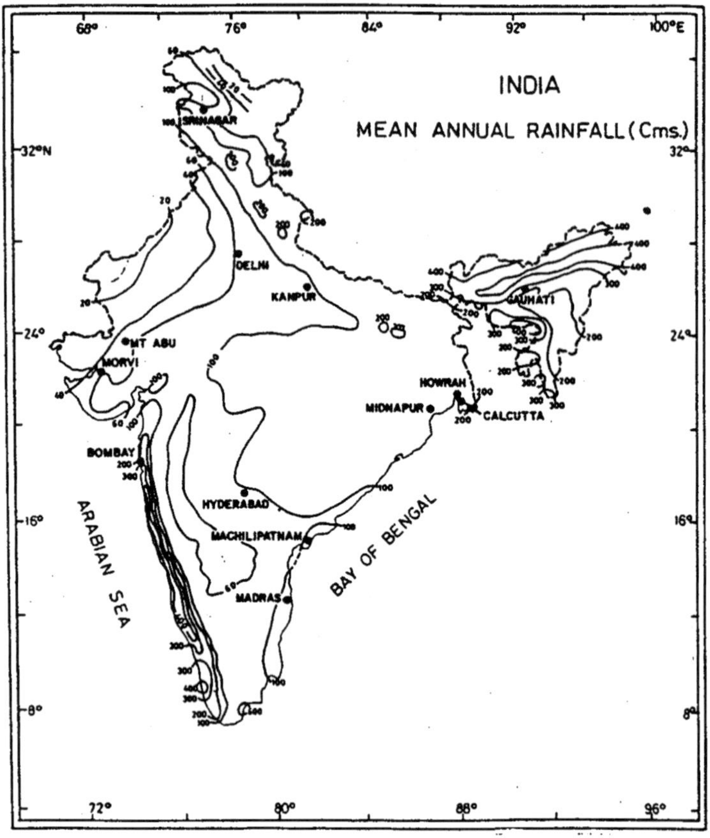

Fig. 1. Mean annual rainfall (Cms) in India.

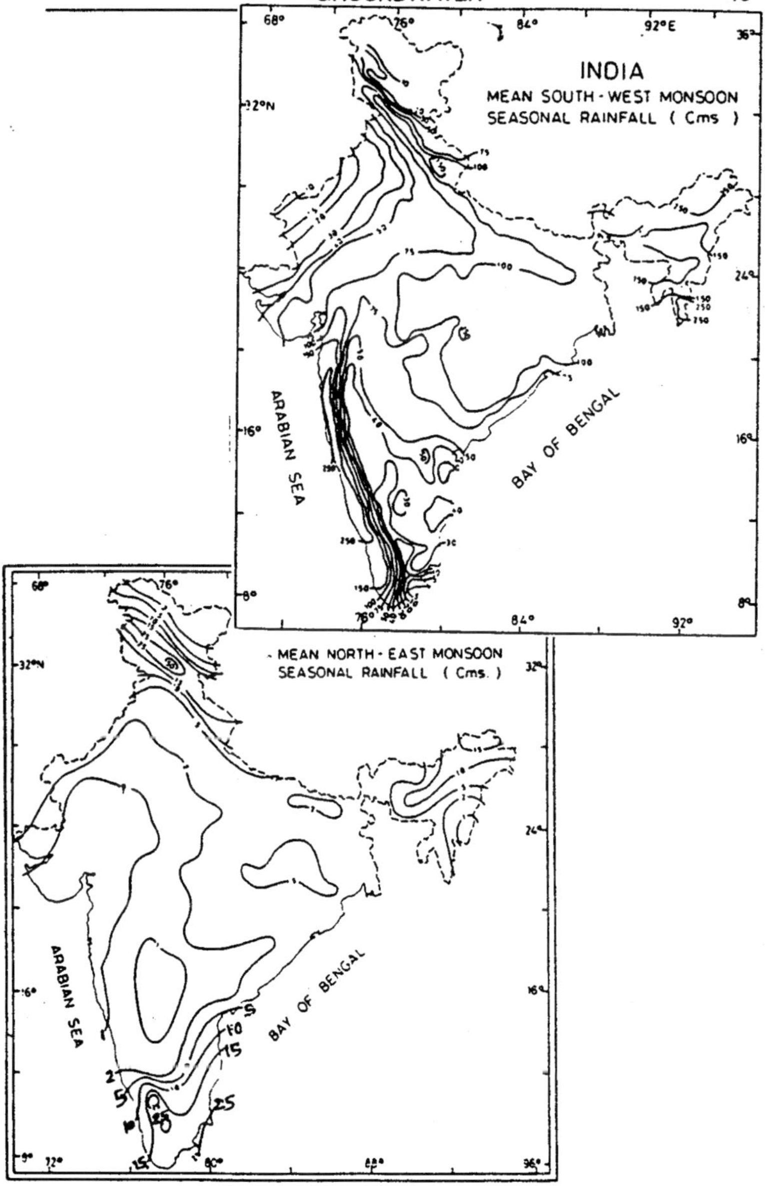

Fig. 2. Seasonal rainfall (cms)

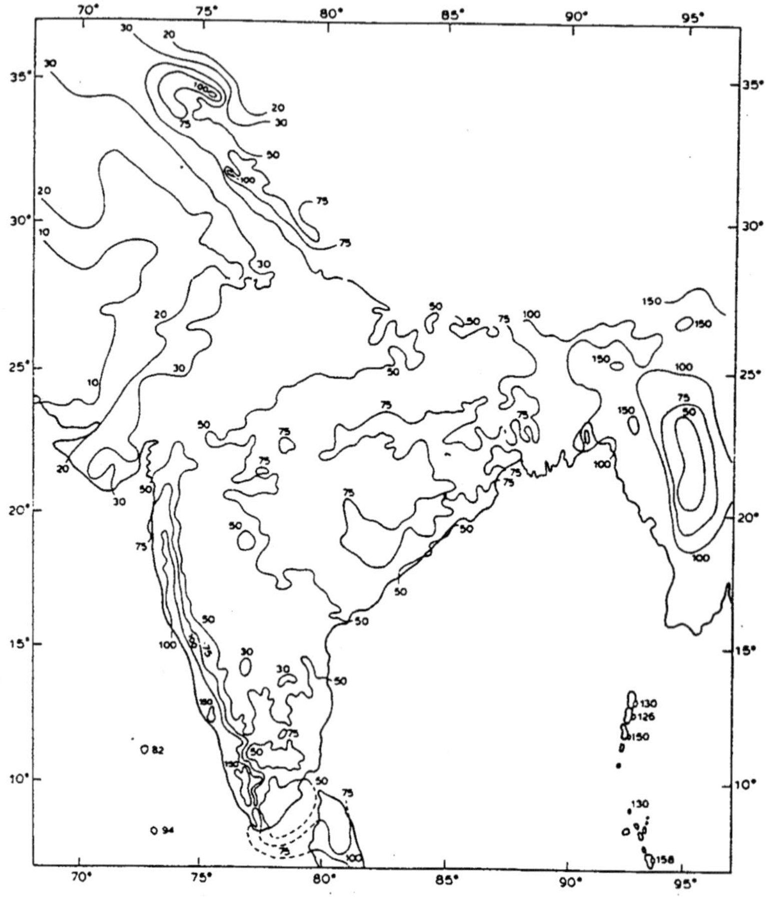

Fig. 3. Mean annual number of rainy days.

Precipitation in winter (Jan-Feb) and in the hot weather period (March-May) is only a very small portion of the annual values and does not contribute much to the hydrology of the region.

Rainfall Variability

Variability of rainfall decreases with increase of the mean rainfall. On the west coast and the eastern region, variability is less than 20% (Fig. 4). East of the Western Ghats, upto the east coast, it rises to 30%. Variability increases from all other sides to the western desert regions, where it reaches 80%.

Seasonally, the coefficient is 50-100% during winter, with 50% near the northern hills and 100% in the central parts. During the hot weather season, it is 30-40% in the north-eastern parts, Kashmir and highlands of southern peninsula and increases to 100% in the central region. During the southwest monsoon period, areas with 75 cms or more rainfall have variability values of 30% or less. It increases to 60% in the western deserts and 100% in the extreme south peninsula. During the post-monsoon period, the coefficient is 40-50% over most of the regions. But, in the northern parts of the sub- continent, it is 60-80% and in the central region, 100%.

Water Resources

The seasonally heavy rainfall has blessed India with three main watersheds-the Himalayan Range with its Karakoram branch in the north, the Vindhyan and Satpura ranges in central India and the Western Ghats on the west coast : all major rivers in India originate in any of these watersheds.

The rivers in India can be broadly classified into the Himalayan group and the Deccan group. The major rivers in the Himalayan group are Indus, Ganga and Brahmaputra (Fig. 5). These rivers are both snow-fed and rainfed and are perennial. About 70% of the total discharge of the Indian rivers is the contribution of the Himalayan rivers.

Major rivers in the Deccan group are Godavari, Krishna, Cauvery, Pennar, Mahanadi, Damodar, Sharavati, Netravati, Bharatapuzha, Periyar, Pamba, Narmada and Tapti. All these rivers are entirely rain-fed and many of them shrink into rivulets when the hot weather season begins. These rivers contribute 30% to the total discharge, about 20% due to the east flowing rivers and 10% from the west flowing.

The total discharge of all rivers in India is estimated to be 1,683,000 million cubic metres per year. But, due to limitations of terrain, non-availability of storage sites, short period of occurrence of rains etc. the utilisable surface water resources is only 67,000 million cubic metres (Raghunath, 1985).

The temporal and seasonal distribution of rainfall creates floods

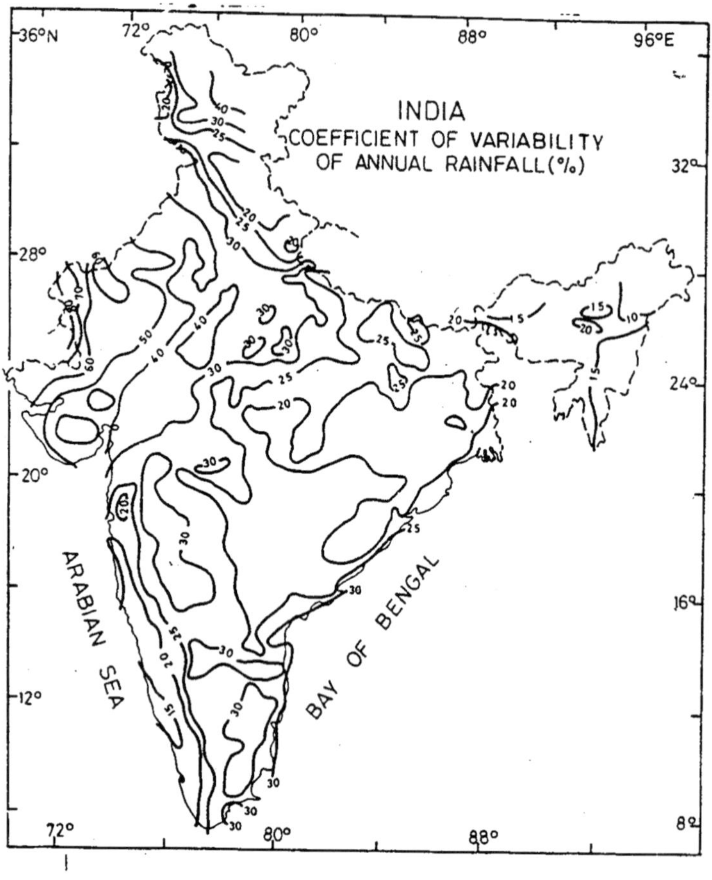

Fig. 4. Coefficient of variability of annual rainfall (%)

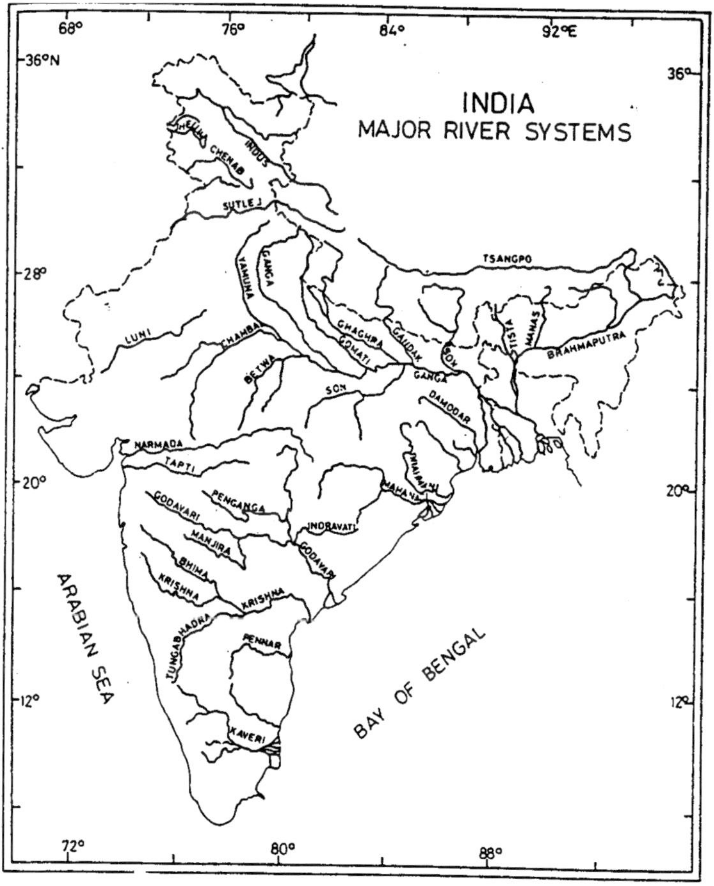

Fig. 5. Major river systems

and droughts in many parts of the country.

Of the total rainfall, 32% is lost by evaporation, 46% flows as runoff and 22% passes through seepage. Of this 22%, only 12% remains in the top soil available for plant growth. Only 35% of the area of India has sufficient water for agriculture. About 41% is arable land with moderate water needs, while the arid and semiarid regions account for 24% of the area (Subrahmanyam & Ram Mohan, 1984). Water supply by precipitation alone is not able to meet the water requirements of individual regions. The index of moisture adequacy, the ratio of the Actual Evapotranspiration to Potential Evapotranspiration is a good measure of the status of the soil in relation to the water need (Subrahmanyam, 1982).

The highest values of this index (80%) are found in north-east India and the Western Ghats region (Fig. 6), indicating rich moisture availability owing to heavy and prolonged precipitation. On isolated locations, the ratio is as high as 100%. The ratio falls to between 40% and 80% in the interior peninsula and central India and below 20% over the north-western region.

In almost all regions in India, including regions of large annual water surpluses, there exist large water deficiencies on an annual basis (Fig. 7). The area of largest water deficiency-more than 120 cms-is found in the northwest (Rajasthan and Punjab). The deficiency decreases towards the east to about 10 cms in the north-eastern regions. Even on the west coast, where rainfall is high, annual deficiencies are about 40 cms. A north-east to south-west increase can also be noticed. In most of the central regions, deficiency is 40 cms. But, in the interior peninsula, it rises to between 80 and 100 cms.

Inadequate amount of rainfall creates periods of acute water shortage termed as droughts, which are severe and frequent in India. No climatic zone in India is free from droughts but their frequency and intensity vary in space and time. There is no periodicity in the occurrence of droughts. Even when the annual rainfall amount is high, the improper distribution can create conditions of drought of varying intensity (Nair, 1987).

Based on the hydrological and agricultural factors, there are three drought-prone areas in India. Of this, 47% is moderate drought-prone area comprising Karnataka, Maharashtra, Gujarat and a narrow strip running from north Madhya Pradesh to southern Uttar Pradesh, Bihar, Southwest Bengal and parts of Rajasthan. The severe drought-prone area covering 42% of the total drought-prone area includes Udaipur, Gwalior and Malwar plateau in the Ghats. The remaining 11% is extreme drought-prone area comprising the desert and the semi-desert regions of western Rajasthan and Kutch.

Excessive and prolonged rainfall in the catchment areas sometimes result in catastrophic floods. On an average, each year 10^5 lakhs hectares of land in India are affected by floods. Floods in the northern

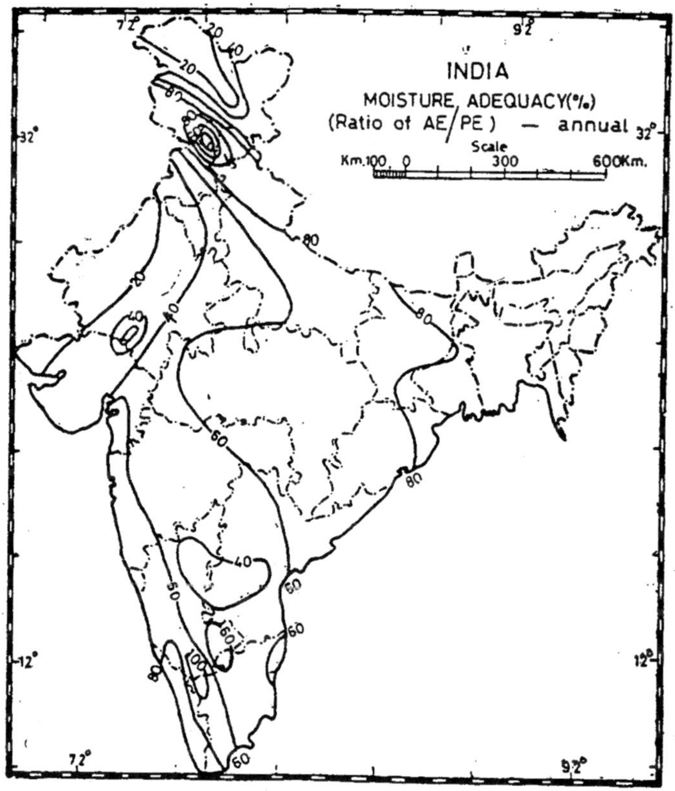

Fig. 6. Index of Moisture adequacy (%)

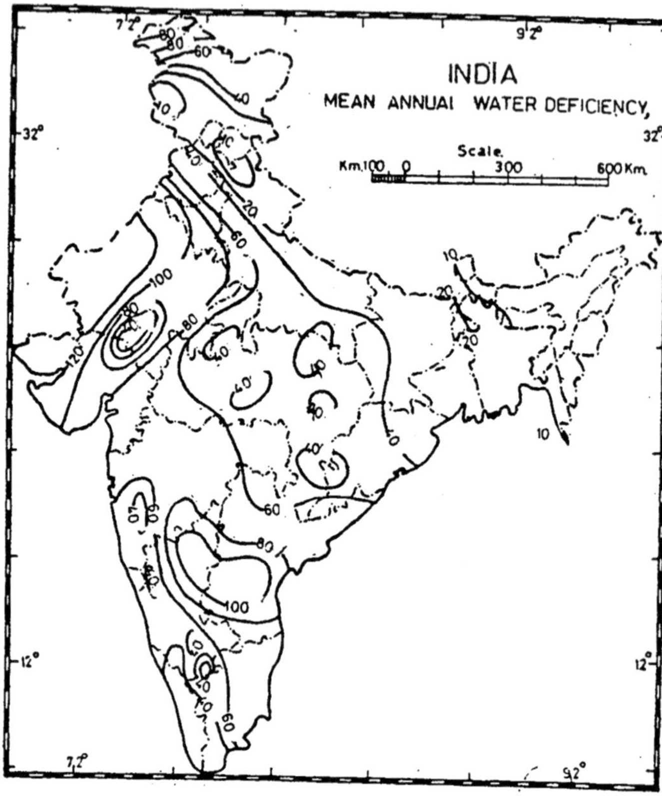

Fig. 7. Mean annual water deficiency (cms)

and eastern parts of India are more severe and frequent than in other parts. All rivers in these regions originate from the Himalayas and in addition to heavy rainfall, they have enough water supply from the snow and glacier melt. Depressions and cyclones create floods of short duration along the Gujarat, Andhra and Tamilnadu coasts. Heavy rainfall in the Western Ghats region often results in small-scale floods on the western side of the Ghats.

Conclusions

It is thus seen that the hydroclimatology of India is mainly determined by the rainfall associated with the monsoon circulation. However, the rainfall produced by the monsoons is not uniformly distributed either in space or in time, which therefore leads to the occurrence of floods and droughts in different parts of the country from time to time. While floods are confined to certain areas based on the distribution of major river systems, droughts occur on account of rainfall failures, following late onset or early withdrawal or both of the monsoons. The rainfall, characteristically exhibits wide variations in magnitude and duration. Consequently, the water budget which is based on water need and water supply by rainfall also shows larger inter-annual fluctuation. Very often, such variations, are of such magnitudes that there are 'climatic shifts' from the normal climatic regimes. Detailed knowledge of such occurrences are necessary in water resources development and water use planning for agricultural or hydrological development.

REFERENCES

Bowen I.S. (1926) The rates of heat losses by conduction and evaporation from any water surface. *Phys. Rev.*, 27, PP. 779-787.

Cummings N.W. (1936) Evaporation from water surface. *Trans. Amer. Geophys. Union*, 17, PP. 507-509.

Ferguson J. (1952) The rate of natural evaporation from shallow ponds. *Aust. Jour. of Sc. Res.*, No. 5, PP. 315-330.

McEwen G.F. (1930) Results of Evaporation Studies. *Scripps. Inst. Oceanog. Tech. Ser.*, 2, PP. 401-415.

Nair Shadananan K. (1987) Hydrometeorological studies of Kerala State in relation to the Western Ghats. Unpublished Ph.D. thesis submitted to the Cochin University of Science and Technology.

Penman H.L. (1956) Evaporation-an Introductory Summary. *Netherlands Jour. Agri. Sci.*, Wagenigen, PP. 9-29.

Pisharoty P.R. (1986) Meteorology for the Indian Farmers, ISRO- PRL-SP-20-86. P. 89.

Raghunath H.M. (1985) Hydrology. Principles, Analysis and Design. Wiley Eastern Ltd. P. 482.

Ram Mohan H.S. (1978) A study of water balance and drought climatology of Tamil Nadu. Unpublished Ph.D. Thesis submitted to Andhra University, Vishakapatnam.

Ram Mohan H.S., Maria Juliet S. and Valsala P. (1984) An agroclimatic approach to the assessment of tropical droughts. *Mausam*, 35, PP. 291-294.

Ram Mohan H.S. and Subrahmanyam V.P. (1984) Application of water balance concepts in agricultural planning and operations in South India. *Nat. Geogr. Jour. of India*, 19, PP. 30-34.

Rao Y.P. (1981) The climate of the Indian Subcontinent. Climates of southern and western Asia. World Survey of Climatology, 19, PP. 67-118.

Richardson B, (1931) Evaporation as a function of insolation. *Trans. ASCE.*, 95, PP. 996-1016.

Slatyer R.O. and McIlroy I.C. (1961) *Practical Microclimatology. UNESCO Pub.*

Subba Rao B. and Subrahmanyam V.P. (1961) Estimation and yields from river basins by a modification of the water balance procedure of Thornthwaite. *Ind. Jour. Met. Geophy.*, 12, PP. 329-334.

Subrahmanyam V.P. (1956) Climatic types of India according to the rational classification of Thornthwaite. *Ind. jour. Met. Geophys.*, 7, PP. 1-12.

Subrahmanyam V.P. (1967) Incidence and spread of continental drought. Reports on WMO/IHD Projects, Report No. 2.

Subrahmanyam V.P. (1982) Water balance and its applications (with special reference to India) Andhra Univ. Press, Waltair PP. 102.

Subrahmanyam V.P. and Ram Mohan H.S. (1980) Classification of monsoon climates and stability of their monsoon regime. *Statistical Climatology*, (Dev. in Atm. Science) No. 13, Elsevier Pub. Co. PP. 335-348.

Subrahmanyam V.P. & Ram Mohan H.S. (1984) Droughts and Agricultural Landuse in India, Pakistan and Bangladesh. Climate and Agricultural Landuse in Monsoon Asia. University of Tokyo Press PP. 183-202.

Subrahmanyam V.P. and Sastri C.V..S. (1971) New techniques in drought analysis. *Annals of Arid Zone*, 10 (2 & 3), PP. 120-135.

Sutton O.G. (1931) Wind structure and Evaporation in a Turbulent Atmosphere, *Proc. Roy Soc. (London)*, Series 146, PP. 701- 722.

Sverdrup H.U. (1937) On the Evaporation from the oceans. *Jour. of Marine Research*, No. 1, PP. 3-14.

Thornthwaite C.W. (1948) An approach towards a rational classification of climate. *Geogr. Rev. 3*, 38(1), PP. 55- 94.

Thornthwaite C.W. and Holzman B. (1942) Measurement of evaporation from land and water surfaces, *Tech Bull. No. 816 U.S. Dep. of Agri.*, Washington D.C.

Thornthwaite C.W. and Mather J.R. (1955) Water budget, its use in irrigation. *Water-Year book of Agriculture U.S. Dept. of Agr.*, Washington DC, PP. 346-358.

3

HYDROLOGIC CYCLE

V. V. S. GURUNADHA RAO

National Geophysical Research Institute
Hyderabad - 500 007

INTRODUCTION

Water differs from most other natural resources by its remarkable property of continuous renewal in the water cycle, the main link being the water exchange between oceans and land. The world ocean is a gigantic evaporator and the main source of freshwater in this natural cycle. This freshwater falls as atmospheric precipitation and is the source of all water flows and water accumulation on land. Another important property of water is that regardless of the number of times it is used by man, the quantity of water on the earth is not reduced, and in the water cycle its purity is restored and it is available for use once again (UNESCO, 1978).

Even though water resources are inexhaustible, the main causes of water problems are firstly that as man's needs increase, the quantity of water required rises proportionately, secondly that, according to forecasts for the first half of the 21st century, this requirement in the world would equal or even exceed the available water resources and thirdly that unused natural water is quickly polluted by the discharge into rivers and other waters of non-purified, or poorly purified industrial, agricultural and domestic waste. The impact of the human activities are becoming evident, not only in the quality and quantity of water in rivers, lakes and groundwater reservoirs, but also in the regimes of inland seas and even in the oceans.

The water balance concept widely used in hydrology may introduce some confusion because the assumed equality of precipitation to the sum of river runoff and evaporation does not always occur in nature. Many river basins, and even sea basins, do not have, strictly speaking exclusive cycles. In general, the smaller the catchment area, the less justification there is for considering its water balance to be complete. Even when examining the water balances of very large territories, such

GROUND WATER
Editor: P. Sankara Pitchaiah, © 1994 Scientific Publishers, Jodhpur

as continents, it is still unreasonable to expect complete balance. Side by side with the term water balance other term hydrologic cycle has gained wide spread acceptance in the recent past. It is envisaged that the term hydrologic cycle be taken to mean the dissemination and movement of all forms of water together with dissolved and suspended matter above and below the surface of the earth. At this juncture the study of water balance should be regarded as a method of investigating the hydrologic cycle.

The term groundwater is usually reserved for the subsurface water that occurs beneath the water table in soils and geological formations that are fully saturated. As the study of groundwater is a interdisciplinary in nature, the study becomes germane to geologists, hydrologists, soil scientists, agricultural engineers, foresters, geographers, ecologists, geotechnical engineers, mining engineers, sanitary engineers and petroleum reservoir analysts. Now-a-days groundwater has become more than a resource like it was decades back. It is an important feature of natural environment, it leads to environmental problems, and may in some cases offer a medium for environmental solutions. It is a part of the hydrologic cycle, and an understanding of its role in this cycle is mandatory if integrated analyses are to be promoted in the consideration of watershed resources, and in the regional as well as global assessment of water resources. Groundwater in various ways acts as a key to the understanding of a wide variety of geological processes viz., generation of earth-quakes, the movement and accumulation of petroleum, the origin of certain types of ore deposits, soil types and land forms.

Water moves through the earth system in an endless cycle that forms the hdyrologic cycle. While moving through, it plays a central role in many of the physical, chemical, and biological processes regulating the earth system making the understanding of the hydrological cycle vital in the study of the environment. Water vapour is the dynamic fluid of the atmospheric heat engine where through evaporation and condensation processes it drives important atmospheric and oceanic circulations and redistributes absorbed solar energy. It is instrumental in controlling the temperature of the planetary system. Through fluvial erosion and sedimentation processes, water contributes to the shape of the land surface. Water is the universal solvent and the medium in which most changes of matter can take place, hence it is the agent of element cycling. Water is also vital for sustaining life.

The role of water in human affairs, the scale of understanding has been made modest beginings with the understanding of surface drainage basins. The gaps in the scientific knowledge of water cycle include the geographical redistribution of water resources due to climate change, the ecological consequences of large-scale water transfers, widespread mining of fossil groundwater, the effect of land use changes on the regional hydrologic cycle, the effect of nonpoint sources of pollution on the quality of surface water as well as groundwater at regional scale, and the possibility of changing regimes of regional floods and droughts. In the early part of the century the main emphasis was aimed at conservation of water resources on river basin scale. The concern was principally with the physics of the hydrologic cycle, omitting the quality aspects and the effects of man were excluded. By the middle of the century the emphasis has been changed to the understanding of laboratory scale physical processes of the hydrologic cycle. Later chemical constituents of water and the relation of water to living things have been included. International Hydrological Decade has raised consciousness about regional and global problems and about man's impact on the hydrologic cycle. Contemporary views of hydrology accept that human activity has become an integral part of the hydrologic cycle and that the quality of water is no less a concern than the quantity. Further the stress has been towards the dynamic interaction of water with the other components of the earth system.

RESERVOIRS AND FLUXES OF WATER

Space and Mantle

The hydrologic cycle is illustrated as a global geophysical process (Figure. 1). The availability of water in space and earth's mantle is known very limitedly about the quantity of water in these two resevoirs, but there is evidence that they possess and both exchange water the primary crustal, ice, atmospheric and ocean reservoirs. The hydrogen in water is being lost to space very slowly through diffusion of water vapour and methane molecules into the upper atmosphere and the subsequent escape of hydrogen atoms freed from these molecules by photochemistry (Eagleson, 1991). This effective loss is, indicated in Figure 1 and the rate is probably of the order of 10 km per annum. The volcanic activity continuously outlets water vapour to the atmosphere and liquid water to the oceans with a rate of 1 km annually. Water also recirculates on geological time scale by the subduction of water containing crustal material in a tectonic hydrologic cycle (Fig. 2).

Oceans and Atmosphere

The abundance of 97.2% of water in the land-ocean-atmosphere system is saline and resides in the oceans. This enormous reservoir forms the main source in the processes of exchange with water on and under the land in the atmosphere on many different space-time scales. The atmosphere supports an average precipitation of the order of 1 m per annum, contains only 0.001% of global water at any moment. This atmospheric water storage is replenished once in every 9 to 10 days on the average through evaporation. Evaporation vapour transport in the atmosphere, condensation, and precipitation are the fundamental mechanisms for distillation and redistribution of earth's freshwater in a climatic hydrologic cycle. This cycle, usually referred to simply as "the hydrologic cycle" plays a major role in the redistribution of

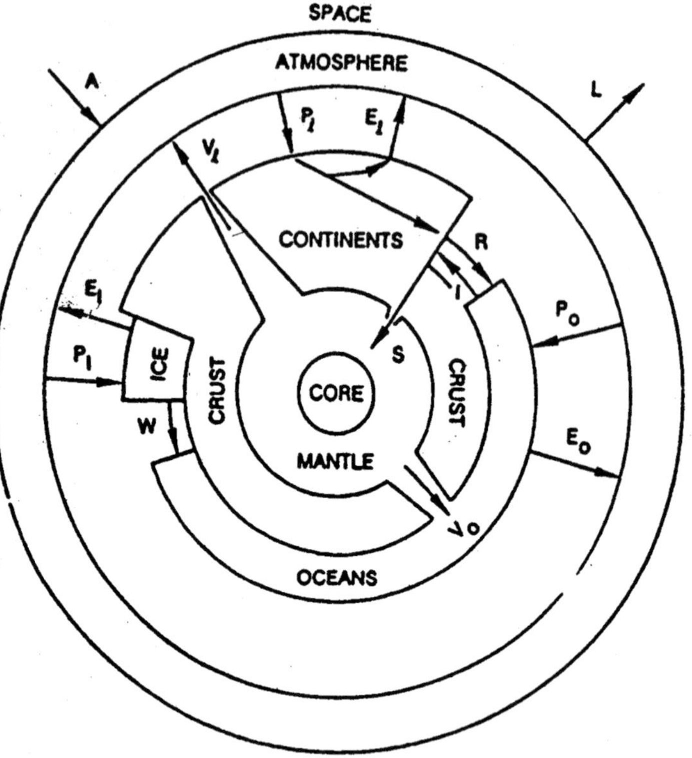

A = Additions of water from space
E_o = Evaporation from oceans
E_i = Evaporation (i.e., sublimation) from ice
E_l = Evapotranspiration from land
I = Intrusion of seawater into continental aquifers
L = Loss of water to space
P_o = Precipitation on oceans
P_i = Precipitation on ice
P_l = Precipitation on land
R = Runoff from continents
S = Subduction of water-containing crust
V_o = Volcanic venting to oceans
V_l = Volcanic venting to atmosphere
W = Wastage of ice sheets to ocean

Fig. 1. Hydrologic cycle as a global geophysical process (from NRC, 1991)

incoming solar energy also. It is illustrated quantitatively on a global scale in Figure 3. About one-half of all solar energy reaching earth's surface is used in evaporation, of which 90% comes from the ocean and rest from water storages on the land. Oceanic precipitation and evaporation have not been observed systematically because of experimental difficulties and global water balances could not be estimated quantitatively.

The space time distribution of storm precipitation is poorly understood in relation to the mechanics of storm genesis even for storms over land where observation is much simpler. The strength of

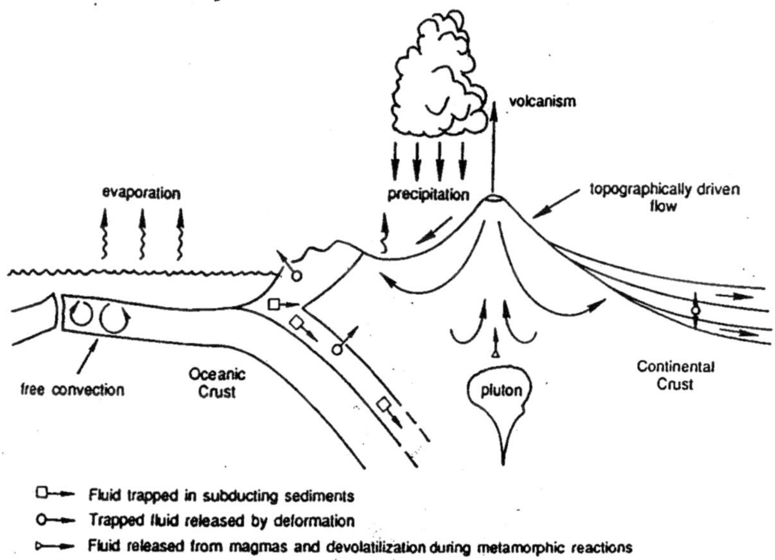

Fig. 2. Tectonic hydrologic cycle (from Foster and Smith, 1990)

evapotranspiration cycle and associated energy redistribution are known to be highly variable with agroclimatic environment. In tropical regions where thermal convection is a dominant atmoshpheric mechanism as much as 50% of local precipitation may be derived from local evaporation. In temperate latitudes where cyclonic precipitation is dominant it may be only 10%.

HYDROLOGIC CYCLE-CONCEPTS AND DEFINITIONS

Conceptualizations

As a global average only about 57% of the precipitation which falls on the land returns directly to the atmosphere without entering the ocean. The run-off as reminder finds its way to the sea primarily through rivers but also through subsurface groundwater movement, and by the calving of icebergs from glaciers and ice shelves. The water may spend some time in one or more natural storage reservoirs in this gravitationally controlled run-off process. The storage reservoirs include snowpack, glaciers, ice sheets, lakes, streams, soils and sediments, vegetation and rock. Evaporation from these reservoirs short circuits the global hydrologic cycle into subcycles with a wide spectrum of scale.

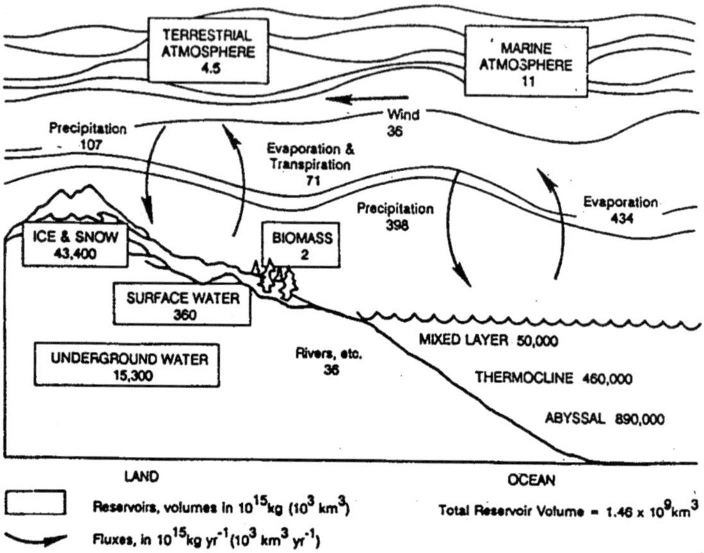

Fig. 3. Climatic hydrologic cycle on global scale. (from NRC, 1982)

The hydrologic processes that may be operative in a watershed are shown in the schematic diagram (Fig. 4) wherein that a watershed must be treated as a combination of both the surface drainage area and the upper part of subsurface soils and geologic formations. In a pot-and-pipeline representation often used in the systems approach to hydrologic modeling. Figure 5 is conceptually clear in that it emphasizes processes and illustrates the flow system concept of the hydrologic cycle, often used when referring to the rates and quantities involved in the various processes (Allan Freeze, 1974).

Subsurface flow processes

A subsurface system as shown in Figures 4 and 5 involves flow through porous media in both the saturated and the unsaturated state. The "saturated (groundwater) zone" occurs below the water table. In this zone the soil or rock pores are all filled with water, and the fluid pressures are greater than atmospheric pressure. In the 'unsaturated (soil moisture) zone' above the water table the soil pores are only partially filled, and water is held in the pores under tension (i.e. fluid pressures are less than atmospheric pressure). The 'water table' is then

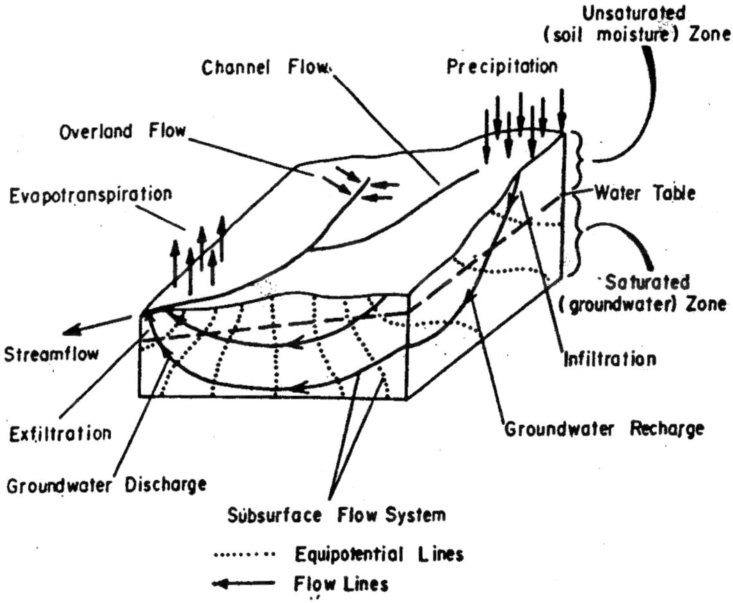

Fig. 4. Hydrologic processes in a watershed (from Allan Freeze, 1974)

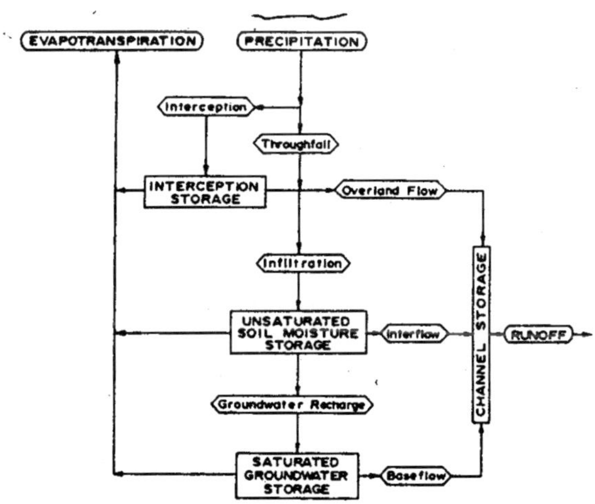

Fig. 5. Rates and quantities in a systems-type conceptualization of the hydrologic cycle (from Allan Freeze, 1974)

defined as that surface on which the fluid pressure is equ· to atmospheric pressure. Its depth at any point can be determined by observing the water level in a shallow well open to the atmosphere.

The process of 'infiltration' refers to the entry into the soil of water made available at the ground surface, together with the associated unsaturated downward flow. 'Exfiltration' is the removal of water from the soil at the ground surface, together with the associated unsaturated upward flow. Also we may define 'groundwater recharge' as the entry into the saturated zone, together with the associated flow away from the water table. 'Groundwater discharge' is the removal of water from the saturated zone across the water table surface, together with the associated flow toward the water table within the s .rated zone. A subsurface flow system may feed water to a stream channel by the processes of subsurface storm flow or groundwater flow. 'Subsurface strom flow' refers to that portion of the stream's lateral inflow that is derived from water that infiltrates through the surface and moves laterally through the upper soil horizons toward the stream channel as unsaturated flow or as shallow perched saturated flow above the main groundwater table. It must enter the stream channel along a 'seepage face', which is that portion of stream bank above the stream flow level and below the line where the water table intersects the bank. 'Groundwater flow' refers to that portion of the stream's lateral inflow that is derived from deep percolation of infiltrated water that enters the permanent saturated groundwater flow system and then discharges into the stream channel.

Rates and quantities

The rates associated with subsurface flow processes are straight forward expressed as 'infiltration rate, exfiltration rate, groundwater recharge rate and groundwater discharge rate'. That part of the stream hydrograph that can be ascribed to lateral inflow from the subsurface strom flow process is called 'interflow', that which can be ascribed to the groundwater flow process is called 'base flow'. 'Overland flow' is the process by which water flows over the land surface toward a stream channel. Flow in the channel is 'stream flow'. The overland flow denote that part of the stream hydrograph that can be ascribed to lateral inflow from the overland flow process. The stream discharge at any point in time at the downstream reach of any channel is termed the 'channel flow'. The total quantity of water that arrives at the outlet of a drainage basin during any specified period is termed the 'runoff'. It is that part of the precipitation that appears as stream flow.

CRUST

The water in the uppermost meter of earth's crust is known as soil moisture and is a key determinant of the state of earth system. With typical residence times of days to months, it provides the driving potential for moisture fluxes in the soil upward as evaporation and as absorption by the roots of plants and downward in recharging groundwater table. It serves the same storage and driving functions for dissolved chemical species. Soil moisture also plays an important role in soil formation, directly through chemical weathering of rock and indirectly through life support of soil biots. Soil moisture also is an active agent as well in the depletion of soil minerals through leaching, and it influences the resistance to soils to erosion. Thus it also has an obvious and direct impact on the growth of vegetation an aspect of the hydrologic cycle that has been intensively studied at the microscale because of its importance to agriculture.

The porous earth and rock materials deep beneath the land surface and oceans constitute a huge water reservoir. It has been estimated that some permeability exists within the crust to depths of 13-20 km. Water filled in these pores is called groundwater, the premeable strata are called aquifers, and the upper limit of the saturated zone approximates to the water table. The general movement of groundwater is the result of topographic relief and the rates of flow are slow, perhaps a few centimeters per day. Depending on how far the groundwater must travel to reach a surface discharge area, water in the shallow zone may remain underground from few hours to more than 100 years. Water at great depth may take tens or even hundreds or thousands of years to pass through the subsurface. The circulation of groundwater is evident at depths from 10 to 15 km, probably due to tectonically induced pressure gradients.

The volume of groundwater in the upper kilometer of the continental crust is an order of magnitude larger than the combined volume of water in all rivers and lakes and is equivalent to the total of all groundwater recharge during the last 150 years. The total volume of groundwater is almost one- fourth of all non-oceanic water on earth. Groundwater discharges are common at topographical lows, which enable streams to flow even during non-rainy periods. Coastal groundwater reservoirs may be replenished locally by saltwater intrusion from the ocean.

ICE

The largest quantity of freshwater occurs as ice in the Antarctic and Greenland ice sheets. This ice with an average residence time of about 10,000 years, participates very slowly in the hydrologic cycle. However, these reservoirs are so large that small percentage changes in volume can cause major changes in sea level on time scales of 100 to 10,000 years. The greatest single item in the water budget on earth, besides the oceans, is the Antarctic Ice sheet. It contains about 64% of all non-oceanic waters. Melting of just the small west Antarctic Ice sheet would raise global sea level by about few meters. The addition or subtraction of ice from smaller ice caps and mountain glaciers, reacting on time scales of 10 to 100 years, may appreciably affect the flow of certain rivers. The melting of ice in these glaciers appears to have caused a third to a half of the sea level rise observed during the last century. The magnitudes of these major water storages and fluxes are shown in Figure 3.

FLUX OF SEDIMENTS

Water moves around on earth's surface and gravity is the main driving force. In the process of shaping the landscape, run-off forms a tree like network of channels into which the flow becomes concentrated. Deficiencies still exist in our understanding of fluid turbulence, which seriously impair our ability to specify the relative rate of transport of various sizes of grains and aggregates in streams and the duration of their storage at various locations within the channel system. This is important not only for its contribution to understanding erosion and deposition but also because many pollutants are moved through the system absorbed on sediment particles.

FLUX OF DISSOLVED SOLIDS

Water is the universal solvent and as it moves through the hydrologic cycle it dissolves and transports in solution solids as well as gases. Rain falling onto the soil surfaces contains various gases and solids in solution. As some of the water infiltrates into soil and moves downward it picks up carbondioxide from into soil, exchanges solutes with soil and rock particles, and becomes less acid. The percolating waters convey their solute load through the groundwater and into streams. This water has a dissolved mineral signature which is dependent upon the subsurface material properties, the flow path, and biological processes which recycle minerals. Knowledge of these solutes and their chemical kinetics can be used in tracer studies of subsurface water flow paths and in understanding the rates of continental degradation and soil formation.

INVOLVEMENT OF BIOTA

Water supports a variety of living organisms and some have major interactions with the hydrologic cycle. The thin soil cover, for instance, is a result of physicochemical weathering by water and supports a vegetal cover which is about 99% of earth's terrestrial biomass. The vegetation is responsible through transpiration for most of the water returned directly to the atmosphere from the land. The removal of vegetation will modify run-off and, in certain climates, may reduce local precpitation due to earth-atmospheric coupling.

The soil also supports a variety of micro-organisms which act upon complex organic materials and reduce them to simpler organic compounds and ultimately to mineral form. Bacteria and fungi are particularly important in the carbon cycle and in regulating the availability of phosphorus, nitrogen and sulfur. Their action results in the production of carbondioxide and the development of humus, the organic detritus of decayed vegetation. Humus affects the infiltration and water holding capacity of the soil and its resistance to erosion. Recent investigations have determined the presence of micro-organisms upto 250 m below ground level. It is interesting to speculate on their natural role, if any, in the hydrologic cycle and on their possible use in the degradation of anthropogenic contaminants. It is well known that oceanic micro-organisms are responsible for roughly one-half of earth's photosynthesis and therefore play a major role in atmospheric chemistry and the chemical quality of precipitation. Wet lands are primary source of atmospheric methane and perform a host of other hydrologic and biogeochemical functions. In the last few centuries the impact of humans has been widely felt on the hydrologic cycle. The interventions of man presently are extended to deserts and mountainous regions, which are fragile environments and about which hydrologic data and understanding are absent. Thus we must hereafter incorporate man as an active component of the hydrologic cycle in all environments.

INTERACTIVE ROLE OF WATER

Water is both actively and passively important in its interactive role to the earth system. In the passive sense, the occurrence and quality of water are sensitive to atmospheric dynamics and composition. The globle distribution of rainfall, snowfall, evaporation and accumulated water, both surface and subsurface control the water availability locally to some extent. Water movement couples the land with the oceans through the solution, entrainment, and transport of minerals and sediments, both liquid water and ice are powerful agents of erosion and, in the long term, they join with plate tectonics in shaping the land surface. In the active sense, changes in land surface properties and vegetation cover many among other things, alter the long-term average run-off water. This would demand a corresponding change in the long-term average convergence of atmospheric moisture and commensurate adjustment in atmospheric dynamics.

DATA ISSUES

Hydrologic processes are highly variable in space and time, and this variability exists at all scales in space and time. Data collection over such a range of scales is highly impossible and expensive, and so hydrologic models usually conceptualize processes based on simple, often homogeneous, models of nature. This forced over simplification is impeding both scientific understanding and management of water resources. Data collection programs provide the basis for understanding hydrologic systems and documenting changes in the regional and global environments. Modeling and data collection are not independent processes. Ideally, each drives and directs the other. Better models illustrate the type and quality of data that are required to test hypotheses. Better data, in turn, permit better and more complete models and new hypotheses. In this process emphasis could be given to the importance of observational and experimental skills.

The hydrologic data are collected for operational purposes as well as for scientific purposes. Improvements in the use of operational data require that special attention be given to the maintenance of continuous long-term data sets of established quality and reliability. The increasing emphasis on global scale hydrology and the increasing importance of satellite and ground based remote sensing lead to use of large volumes of data that are collected by many different agencies. An information management system is needed that would allow searching many data bases and intergrating data collected at different scales and by different agencies. Effective use of remote sensing data is now difficult for many hydrologists, because the interpretation often depends on a detailed knowledge of sensor characteristics and electromagnetic properties of the surface and atmosphere. Hydrologic data products should be made accessible in a form such that scientists who are not remote sensing experts can easily use the information derived.

SCIENTIFIC PRIORITIES

In combination with components of the hydrologic cycle, aqueous geochemistry is the key to understand many of the pathways of water through soil and rock, to reveal historical states having value to climate research, and to reconstruct the erosion history of continents. Together with the physics of flow in geologic media aquatic chemistry and microbiology will reveal solute transformations, biogeochemical functioning and the mechanisms for both contamination and purification of soils and water. Water is the basis for ecosystem structure, and many ecosystems are active participants in the hydrologic cycle. Understanding these interactions between ecosystems and the hydrologic cycle is essential for interpreting, forecasting, and even ameliorating global climate change.

In many guises *throughout* the hydrologic assessment we encounter questions concerning the quantitative relationship between the same process occuring at disparate spatial or temporal scales. Most often, these are problems of complex aggregation that are confounding our attempts to quantify predictions of large-scale hydrologic processes. It arises in attempting to extend tracer tests, carried out upto depths of 10 to 50 m in an aquifer, to prediction of solute transport over distances of hundreds of meters or kilometers. It mainly occurs in extrapolating measurements of medium properties in a small number of deep boreholes to characterize fluid fluxes of crustal depth. The inverse problem, disaggregating conditions at large scale to obtain small scale information, arises commonly in the parameterization of subgrid-scale process in climate models and in inferring the subpixel properties of remote sensor images. Our knowledge of the time and space distributions of rainfall, soil moisture, groundwater recharge, and evapotranspiration is very much inadequate, partly because historical data bases are point measurements from which we have attempted extrapolation to large scale fields. The knowledge of their variability and the sensitivity of local and regional climates to alterations in land surface properties is particularly poor. One should go for multidisciplinary experiments to come over these problems.

REFERENCES

Allan Freeze, R. 1974 Streamflow Generation. *Reviews of Geophysics and Space Physics*, Vol. 12, No. 4, p. 627-647.

Engleson, P.S. 1991 Hydrologic Science : A distinct Geoscience. *Reviews of Geophysics*, Vol. 29, No. 2, p. 237-248.

Foster, C. and Smith, J.L. 1990 Fluid flow in tectonic regimes, in Fluids in Tectonically Active regimes of the continental crust, *MAC Short Course* Vol. 18 (edited by B.E. Nesbitt) Mineralogical Association of Canada, Vancouver, B.D., p. 1-47.

National Research Council, 1982 *Scientific basis of water resources management*. National Academy Press, Washington, D.C. pp. 127.

National Research Council, 1991 *Opportunities in the Hydrologic Sciences*. National Academy Press, Washington, D.C.

UNESCO, 1978 World water balance and water resources of the earth. Studies and reports in hydrology 25.

4

HYDROCHEMISTRY

Dr. P. SANKARA PITCHAIAH

Dept. of Geology, Nagarjuna University,
Nagarjuna Nagar - 522 510 (A.P.)

This chapter presents the various physical and chemical properties those reflect hydrochemistry of water. It also explains the methods to represent the analytical data. As the chemical properties of water influence the chemical reactions that take place in aqueous solution, it is pertinent to discuss the properties of water before describing the other aspects.

Water is anomalous in its physical and chemical properties (Table -1). Majority of these anomalies reflect the ability of water molecules to cling tenaciously to one another and to other substances.

Table- 1 : Some physical properties of H_2O

Property	H_2O
Molecular weight	18.0151
MP/°C	0.00
BP/°C	100.00
Temperature of maximum density/°C	3.98
Maximum density/g cm^{-3}	1.0000
Density (25°)/g cm^{-3}	0.997 01
Vapour pressure/mmHg	23.75
Viscosity/centipoise	0.8903
Dielectric constant E	78.39
Electrical conductivity (20°C)/ohm^{-1} cm^{-1}	5.7 x 10-8
Ionization constant $[H^+][OH^-]/mol^2\ l^{-2}$	1.008×10^{-14}
Ionic dissociation constant K=$[H^+][OH-]/[H_2O]/mol\ l^-$	1.821 x 10-16
Heat of ionization/kj mol^{-1}	56.27
H^o_f/kj mol^{-1}	-285.35
G^o_f/kj mol^{-1}	-237.19

GROUND WATER
Editor: P. Sankara Pitchaiah, © 1994 Scientific Publishers, Jodhpur

Fig. 1 summarises a few properties and geometrical features of the water molecule and the hydrogen bond in water. The water molecule (Fig. 2) can be represented as a sort of truncated jack contained in a distorted tube. Two arms of jack contain protons, while other two arms formed by P-orbitals. The development of positive and negative charges on individual water molecules result the large dipole movement. The strength of the Hydrogen bond in water arises from the interaction of these charged polar regions and is further fortifies by several possible resonance configurations.

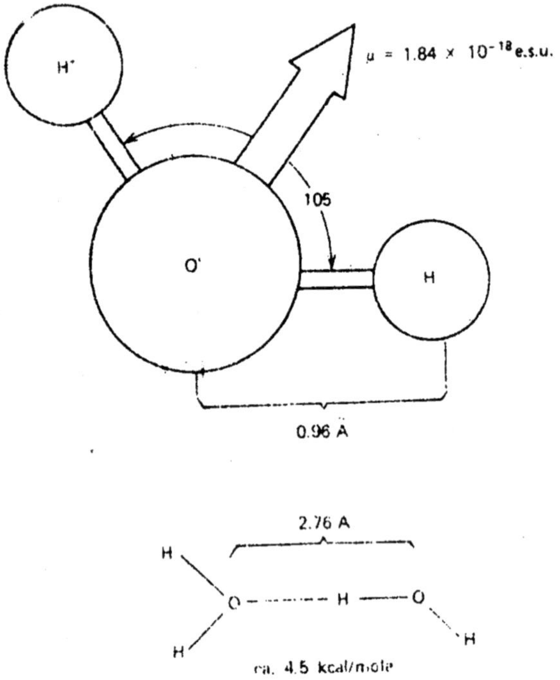

Fig. 1. Schematic summary of the structural features of the water molecule and the hydrogen bond (from Horne, 1969).

Physical properties

Water is a volatile mobile liquid. In the gas phase the water molecule has a bond angle of 104.5° and an inter-atomic distance of 95.7 pm. The dipole movement is 1.84 D. The high boiling point is notable as is the temperature of maximum density and its marked dependence on the isotopic composition of water. The high dielectric constant and measurable ionic dissociation equilibrium are also unusual and important properties. The ionic mobilities in water are

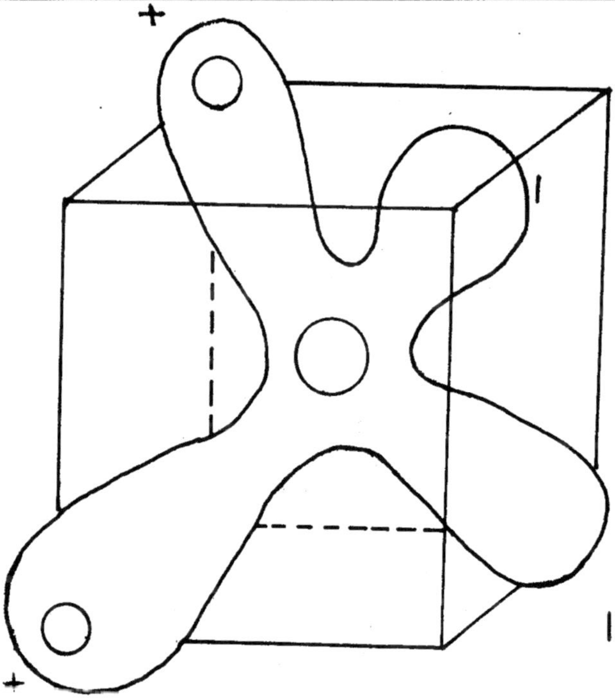

Fig. 2. Electronic cloud of the water molecule (from Horne, 1969).

abnormally high. This has been ascribed to a proton switch and reorientation mechanism involving the ions and chains of hydrogen bonded solvent molecules. Other properties which show the influence of hydrogen bonding are the high heat and entropy of vaporisation, high surface tension and relatively high viscosity. The structural nature of liquid water in which the molecules are linked to a small number of neighbours by H-bonds also accounts for its anomalously low density.

Units of Measurement

Chemical analysis can be measured in several ways. The weight per weight units (ppm : parts per million and ppb : parts per billion) are no longer commonly used. Today, weight per volume units are commonly used. Common units are milligrams per litre (mg/1) and micrograms per litre (μg/1). When the chemical behaviour of the solute is being considered equivalent weight units are useful.

$$\text{Milli equivalents per litre (meq/1)} = \frac{\text{Concentration in mg/1}}{\text{equivalent weight}}$$

$$\text{Equivalent weight} = \frac{\text{formula weight}}{\text{Magnitude of charge on anion or cation}}$$

Dissolved species like silica, which is not ionic cannot be expressed in meq/1.

Molarity is the number of moles of solute present in one litre of solution

$$\text{Molarity} = \frac{\text{Weight of solute in grams}}{\text{Gram formula weight of solute x Volume of solution in liters}}$$

Normality is the number of equivalent weights present in one litre of solution

$$\text{Normality} = \frac{\text{Weight of solute in grams}}{\text{Gram equivalent weight of solute x Volume of solution in liters}}$$

Molarity is the number of moles of solute present in one kilogram of solvent.

$$\text{Molality} = \frac{\text{Weight of solute in grams}}{\text{Gram formula weight x Weight of solvent in kg}}$$

The detailed study of water chemistry is far beyond the scope of this chapter. It is concentrated on the aspects of solubility of gases and liquids in dilute aqueous solutions. All the reactions are assumed to take place at 25°C and a pressure of 1 atmosphere.

PROPERTIES OF AQUEOUS SOLUTIONS

Law of Mass Action

It describes the rate at which substances react together and suggest that the reaction will strive to reach equilibrium. In the general reaction where 'a' moles of A react with 'b' moles of B to form 'c' moles of C and 'd' moles of D, written as

$$aA + bB = cC + dD$$

At equilibrium, the equilibrium constant K is given by

$$K = \alpha C^c \ \alpha D^d / \alpha A^a \ \alpha B^b$$

Where α denotes thermodynamic concentration.

The equilibrium constant for a slightly soluble salt is termed the solubility product K_{sp}. For example the experimentally determined K_{sp} for the reaction

$$AgCl = Ag^{+} + Cl^{-} \text{ is } 10^{-9.8}$$

Chemical Properties

Water has unique properties responsible for the interactions with the material which water contacts throughout the hydrologic cycle. These properties include solvent action, dissociation and transparency.

Solvent Action

Of all the naturally occurring liquids, water is the most effective solvent. The polarity of charge in the water molecule favors the disruption of ionic crystals by reducing the inter ionic attractive forces. Sodium chloride and salts of potassium are readily dissolved in water. Water can also solvate separated ions. Salvation entails the surrounding of charged solute particles by forces. Cations, positively charged ions, are effectively solvent molecules in response to attractive forces. They are effectively solved by compounds of elements in the first row of the periodic table that have unshared electron pairs. Anions, negatively charged ions, are easily solvated by a solvent in which a strongly electronegative element such as Oxygen is bonded in hydrogen. Water satisfies both requirements and consequently is effective in solvating both cations and anions.

Dissociation

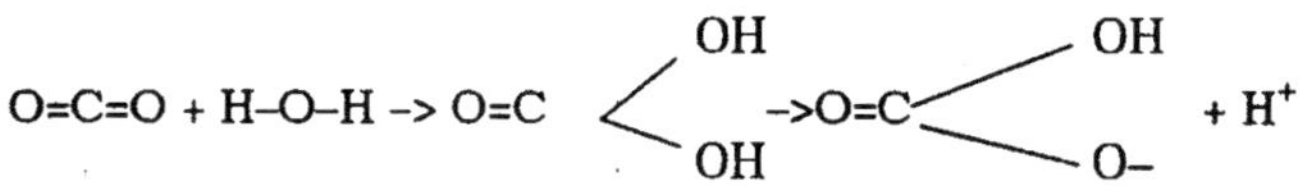

Water is not only a solvent for other substances but is itself capable of dissociating into two charged ions. Hence, water acts as both a base and an acid. An example of this process is the solution of carbon dioxide in water.

Transparency : Solar radiation arriving at the earth's surface is characterized by wavelengths between 0.3 and 1.3 μm. Nearly all radiations outside the range of visible wavelengths greater than 0.8 μm (infrared) results in significant heat transfer to water. The various spectral bands of visible radiants are observed differentially. The red component penetrates no more than 4m of water while some of the blue component penetrates 70m or more. The intensity of a column of pure water is given by beer's law.

$$q_z = q_o e^{-\eta z}$$

Where q_z is the intensity of depth Z, q_o is the intensity of light at the water surface, and n is an extinction coefficient for a specific wavelength of light. Approximately 50 percent of the total incident light is scattered or transformed into heat within the first meter of water.

The ability of water to transmit light is critical to the organisms living in the aquatic environment.

Common Ion Effect

The concentration of a particular ion in an ionic reaction can be increased by the addition of a compound which produces that ion upon dissociation. The particular ion is thus derived from the compound already in solution and also from the added reagent, hence the name 'common ion'. The dissociation of a weak acid/base will decrease in the presence of a substance having common ion. It is called 'common ion effect'.

For example, the solubility of AgCl in pure water is 1.26×10^{-5} mole, while in a 0.1 molal solution of NaCl, it is only 1.58×10^{-9} mole.

Activity Coefficient

The effective concentration of a dissolved substance is referred to as its activity. In dilute solutions the molality is vertually equal to the activity; for more concentrated solutions a correction factor must be used. The correction factor (activity coefficient) that provides a connection between activity and molality of individual ionic species is given by

$$r_i = \alpha_i / mi$$

Where r_i is the activity coefficient of component i and m is the molality.

In summary, to convert from standard chemical analysis to activities,

(a) Convert reported units to molality, mi,

(b) calculate the ionic strength I,

(c) determine individual ion activity coefficients, ri, and

(d) multiply mi by ri to obtain individual ion activities, αi.

The ion activity product Kiap is determined from this information for every reaction for which thermochemical data are available and compared with the reaction.

$$\Delta G^oR = \Delta G^of \text{ products} - \Delta G^of \text{ reactants}$$

When ΔG^oR is zero, no net chemical work is obtainable; the system is in equilibrium. If ΔG^oR is negative, the process may proceed spontaneously; that a positive value means that work (energy) must be put into the system for the reaction to proceed.

The equilibrium constant can be calculated from the Gibbs free energy of reaction according to the relation :

$$\Delta G^o{}_R = -RT/n \text{ Keq}$$

Where R is the gas constant and T is the temperature in degrees Kelvin.

The principal significance of the equilibrium constant in hydrogen chemistry is that it provides a means of calculating the departure from equilibrium between the aqueous phase and coexisting mineral solid phases. The departure is determined by comparison of the thermodynamic equilibrium constant, Keq with the corresponding ion activity product K_{iap}. If the ratio K_{iap} : Keq is less than unity the mineral may be actively dissolving, ratio is unity the water is saturated; a ratio greater than unity implies supersaturation.

Oxidation potential

Many chemical elements are capable of existing in more than one valence state. Thus, for instance, iron may exist as metallic iron with no net charge, as the ferrous ion with a net charge of plus 2, or as the ferric ion which has a charge of plus 3. The stability of an element in a particular oxidation state depends on the energy change involved in adding or removing electrons. A quantitative measure, for this energy change is provided by the factor known as the oxidation potential (also known a redox potential or oxidation-reduction potential).

The oxidation potential of an aqueous solution is called the Eh. The sign of the potential is positive if the reaction is oxidizing and negative if it is reducing. The absolute value of E^o is a measure of the oxidizing or reducing tendency.

The oxidation potential of a reaction is given by the 'Nernst equation

$$EW = E_o + \frac{RT}{nF} \ln K^o$$

Where R is the gas constant, 0.00199 Kcal / (mole.^{o}K)

T is the temperature (degrees Kelvin)

D is the Faraday constant, 23.1 Kcal/volt

n is the number of electrons

If the pH and Eh of an aqueous solution are known, the stability of minerals in contact with the water may be determined. This relationship is best represented as an Eh-pH diagram (Krasukopf, 1967). Water itself, is stable only in a certain part of the Eh- pH field. Water in nature at near surface environments is usually between pH 4 and pH 9, although values that are more acid or more basic can occur.

Carbonate Equilibrium

Neutral water exposed to CO_2 in the atmosphere will dissolve CO_2 equal to the partial pressure. The CO_2 will react with H_2O to form H_2CO_3, a weak acid, and the resulting solution have a pH of about 5.7. Soil CO_2 from organic decomposition is another source. As calcite and dolomite are soluble in acid solution, even rain water will dissolve carbonate rocks. Likewise, a change in pH can result in a precipitation of $CaCO_3$ from a solution that was at equilibrium prior to the pH shift. The equilibrium constants for the various reactions in carbonate equilibrium are given in table 2.

Table- 2 : Carbonate equilibrium constants at 25°C

1. $CaCO_3\ Ca^{++} = \rightarrow CO^3\ CaCo_3 = Ca^{++} + Co_3^{=}$

$$KCaCO_3 = \frac{\alpha\, Ca^{++}\ \alpha\, CO_3}{\alpha\, CaCO_3} = 10^{-8.35}$$

2. $H_2CO_3 = H^+ + HCO_3^-$

$$K_{H2}CO_3 = \frac{\alpha\, H^+\ \alpha\, HCO_3^-}{\alpha\, H_2CO_3} = 10^{-6.4}$$

3. $HCO_3^- = H^+ + CO_3^=$

$$KHCO_3 = \frac{\alpha\, H^+\ \alpha\, CO3^-}{\alpha\, HCO_3^-} = 10^{-10.3}$$

4. $H_2O + CO_2 = H_2CO_3$

$$KCO_2 = \frac{\alpha\, H2CO_3}{PCO_2} = 10^{-1.5}$$

Surface Phenomena

The ionic species present in aqueous solution can react with surfaces of the rocks and solid particles. The surface area of a soil or sediment increases with decreasing grain size, so that the clays are most reactive. There are two separate processes : ion exchange, in which an ion in the mineral lattice is replaced by one of the ions in the aqueous solution; and absorption, in which the solid surface attracts and retains a layer of ions from the solution. These processes may be important in determining the chemical composition of natural groundwaters.

REPRESENTATION OF HYDROGEOCHEMICAL DATA

Both surface and groundwater contain many ions in varying amounts. By means of analytical methods concentration of these ions can be determined. To study the variation of these ions in waters of particular area or to compare with other areas some method of representation is necessary. There have been a great number of graphical forms to represent the hydrogeochemical data. Different types of such graphs are presented below.

(a) Distribution Maps

These maps are normally used for representing the total dissolved solids or electrical conductivity (fig. 3). Such maps provide useful

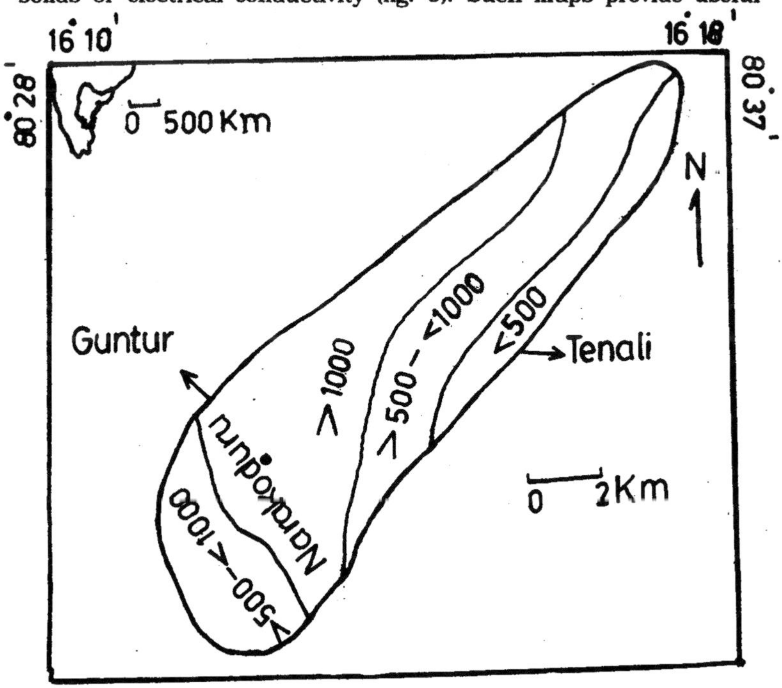

Fig. 3. Spatial variation for TDS (Inset : the study area)

preliminary information about quality of the water. Vector diagrams have little merit.

(b) Hydrochemical sections

The overall chemical trends of chemical concentration are represented and understood using hydro-chemical sections on which a number of parameters can be shown (Lloyd and Heathcote, 1985). Such diagrams (See Fig. 6.4, p. 118, Lloyd and Heathcote) depict associated changes in parameters values and aid interpreting the controlling hydrochemical processes.

(c) Vertical bar graphs

Groundwater analysis (on a bar graph) is shown in fig. 4. Each analysis appears as a vertical bar with a height proportional to the total

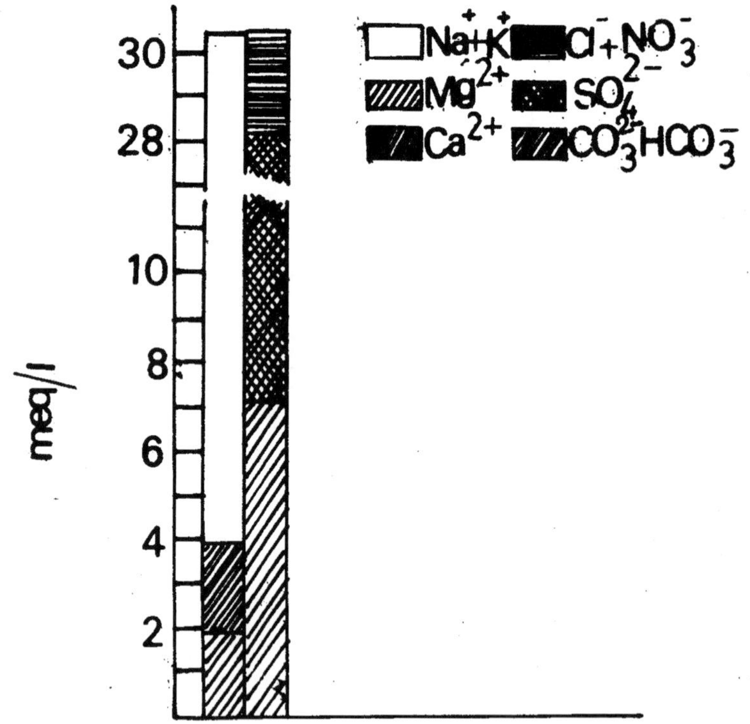

Fig. 4. Bar Chart

concentration of anions or cations.

(d) Circular and radial diagrams

Circular diagrams are constructed with their radii proportional to the total ionic concentration (fig. 5). Sectors within a circle represent fractions of the different ions, expressed as milliequivalents per litre.

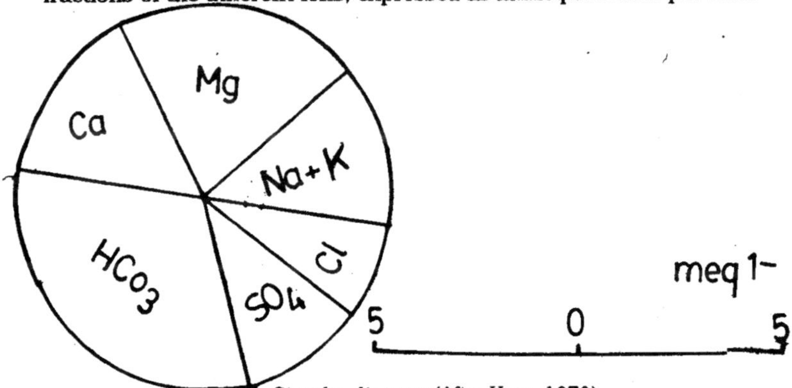

Fig. 5. Circular diagram (After Hem, 1970).

On the radial diagrams (fig. 6) the ions are scaled and the intercepts are joined to represent the water. These diagrams have limited value.

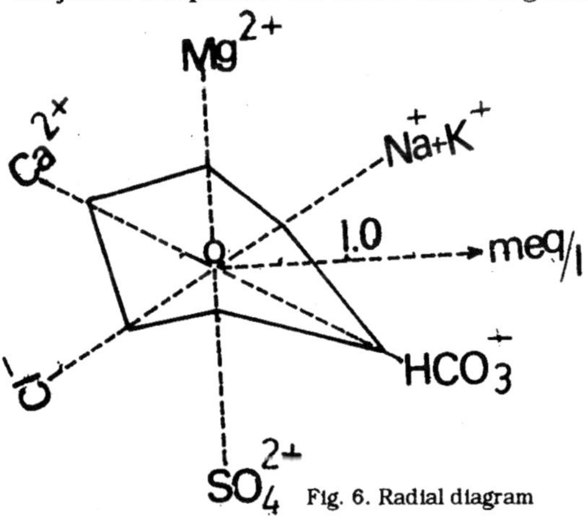

Fig. 6. Radial diagram

(e) Pattern diagrams

Stiff (1951) devised these diagrams to represent chemical analyses in distinctive graphical shapes by plotting ions on parallel axes as shown in fig. 7.

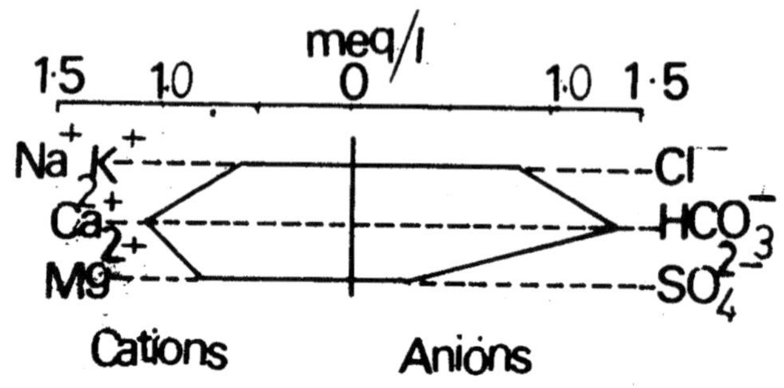

Fig. 7. Pattern diagram (After Stiff, 1951)

(f) Semilogarithmic diagrams

These diagrams were developed by Schoellar (1962) to represent major ion analyses and to demonstrate different hydrochemical water types on the same diagram (fig. 8).

(g) X-Y plots

These plots are the most simple and obvious initial approach to the interpretation of hydrochemical data. A variation of the X-Y plot is the 'dilution diagram' in which the groundwaters in a system are compared with respect to end point waters. Where waters fall on a dilution line a possible mixing percentage can be calculated (Lloyd and Heathcote, 1985).

(h) Trilinear diagrams

Among the various trilinear methods of Plotting (Palmer, 1911; Immense and Harrington, 1913; Hill. 1940; Piper 1944) Piper's diagram (Fig. 9.) has been extensively used to understand problems of geochemical evolution of groundwaters.

Hill (1940) was proposed the trilinear diagram for the first time and refined by Piper (1944).

The Piper trilinear diagram (Fig. 9) combines three distinct fields for plotting, two triangular fields at the lower left and lower right respectively, and an intervening diamond shaped field.

Each apex of a triangle represents a 100 percent concentration of one of three chemical constituents. Major ions are plotted in the two base triangles of the diagram as cation and anion percentages of

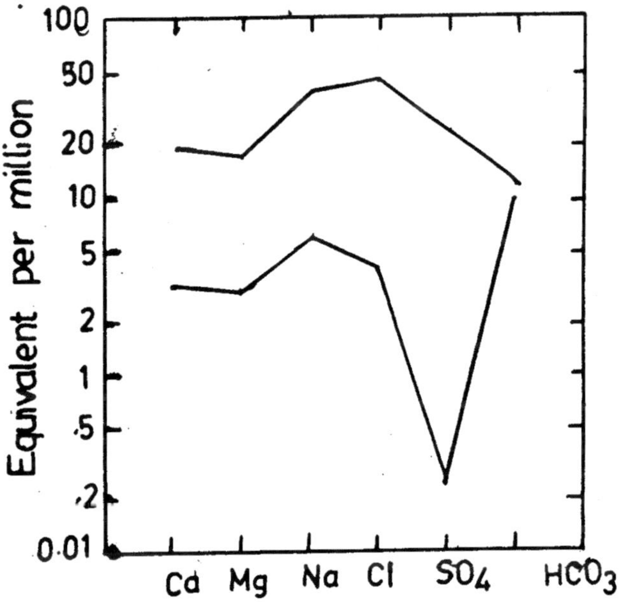

Fig. 8. Schoeller Nomograph

milliequivalents per litre. Total cations and total anions are each considered as 100 percent. The respective cation and anion locations for an analysis are projected into the diamond shaped area which represents the total ion relationship. A circle can be drawn at this point with its area proportional to the dissolved solids. The diagram is also useful in verifying water masses identified by a factor analysis.

All the trilinear systems omit silica from consideration. Because of the importance of silicates in rocks, consideration of silica in water is often desirable where correlations between water analyses and kinds of source rocks are to be made (Hem, 1959). Lloyd and Heathcote (1985) observed following facts about the Pipers trilinear diagram.

(1) The Piper diagram allows comparisons to be made between a large number of analyses but has the drawbacks that all trilinear diagrams have not portraying actual ion concentrations.

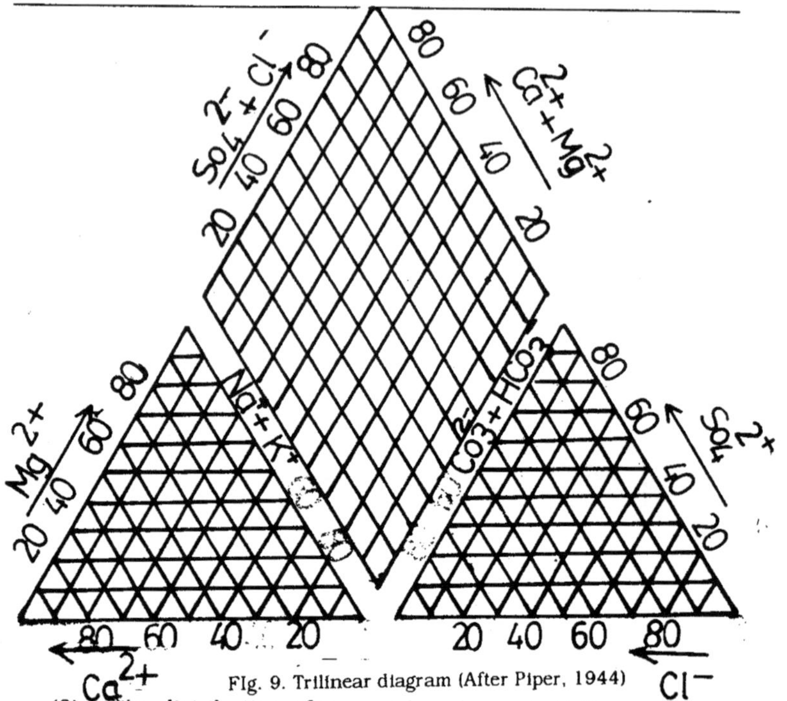

Fig. 9. Trilinear diagram (After Piper, 1944)

(2) The distribution of ions within the main field rectangle is unsystematic in hydrochemical process terms so that the diagram lacks a certain logic. Further, the large amount of line work is a hindrance.

Hydrochemical Facies

The term hydrochemical facies is used to describe the bodies of groundwater, in an equifer, that differ in their chemical composition (Fetter, 1988). The facies are a function of the lithology, solution kinetics, and flow patterns of the aquifer (Back, 1960, 1966). Hydrochemical facies can be classified on the basis of the dominant ions in the facies by means of the trilinear diagram (fig. 10).

Distinct groundwater quality types can be quickly discriminated by plotting groundwater concentrations in certain sub-areas of the diamond-shaped field, as indicated by figure 10 and the following explanation (Piper, 1953)

Area 1 : Alkaline earths exceed alkalies

Area 2 : Alkalies exceed alkaline earths

Area 3 : Weak acids exceed strong acids

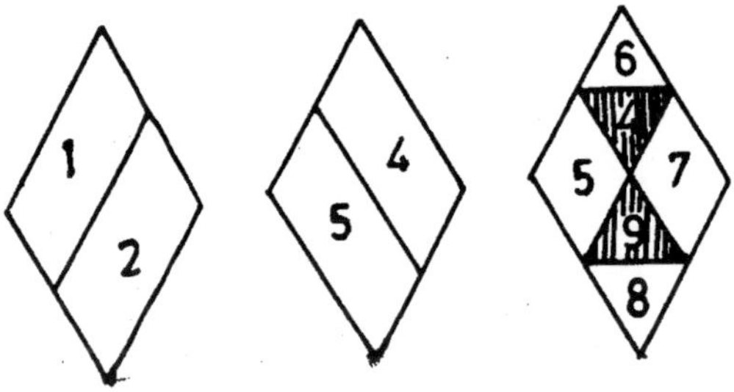

Fig. 10. Subdivisions of the diamond - shaped field

Area 4 : Strong acids exceed weak acids

Area 5 : Carbonate hardness (secondary alkalinity) exceed 50% that is, chemical properties of groundwater are dominated by alkaline earths and weak acids.

Area 6 : Non-carbonate hardness (secondary salinty) exceeds 50 percent

Area 7 : Non-carbonate alkali (primary salinity) exceeds 50 percent - that is, chemical properties are dominated by alkalies and acids

Area 8 : Carbonate alkali (primary alkalinity) exceeds 50 percent

Area 9 : No one cation - anion pair exceeds 50 percent

Mixtures of two groundwaters in all proportions, if all products remain in solution, plot in the three fields on the respective straight lines that join the points representing the respective chemical characters of the two groundwaters mixed (Piper, 1953).

Zekai Sen and Al-Dakheel (1986) found following short comings :

(1) All the cations and anions must be reduced to three groups. For instance, one must lump among cations, potassium and sodium together and among anions fluorids and nitrate with chloride and bicarbonate with carbonate. In-doing so the individual effects of these lumped anions and cations cannot be accurately assessed.

(2) Proportional amounts of anions and cations of two or more different water samples plot as a single point on trilinear graphs.

An alternative diagram to that of Piper has been devised by Durov (1948). In this diagram the cations and anions together total 100 percent. The cation and anion values are plotted in the appropriate triangles and projected into the square main field (fig. 11).

Burdon and Mazloum (1958) and Lloyd (1965) developed an expanded version of the Durov diagram (Fig. 12). Here the cation and anion triangles are recognized and are separated along the 25 percent

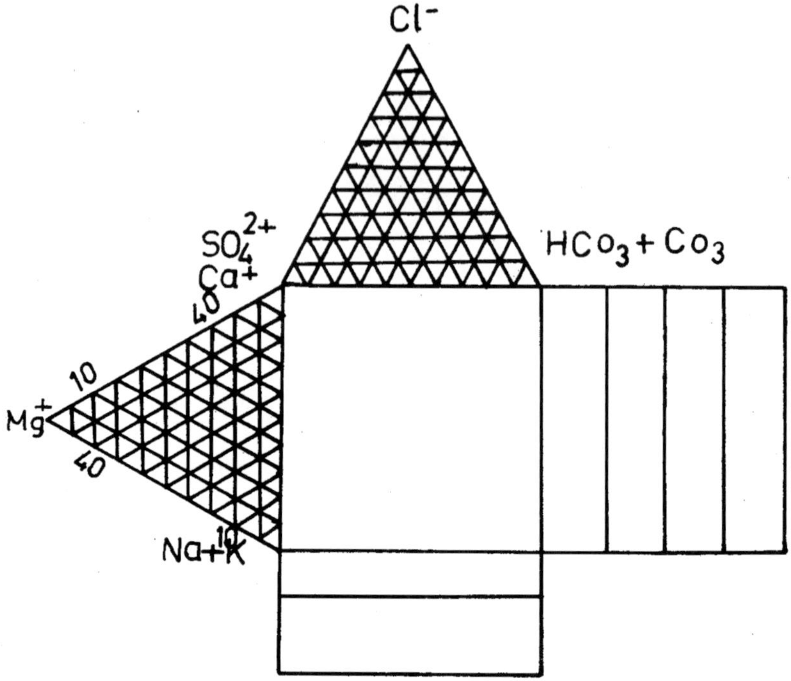

Fig. 11. Durov diagram

axes so that the main field is conveniently divided.

The significance of the nine fields on the expanded Durov diagram (Figs. 12 & 13) can be discussed. (Lloyd and Heathcote, 1985).

1. HCO_3^- and Ca^{+2} dominant, frequently indicates recharging waters in limestone, sandstone and many other aquifers.

2. HCO_3^- dominant and Mg^{+2} dominant or cations indiscriminant, with Mg^{2+} dominant or Ca^{2+} and Mg^{+2} important, indicates waters often associated with dolomites, where Ca^{2+} and Na^+ are important, partial ion exchange may be indicated.

3. HCO_3^- and Na^+ dominant, normally indicates ion-exchanged waters although the generation of CO_2 at depth can produce HCO_3^- where Na^+ is dominant under certain circumstances (Winograd and Farlekas, 1974).

4. $CO_4^{=}$ dominant or anions indiscriminant and Ca^{+2} dominant, Ca^{2+} and SO_4^{-2} dominant frequently indicates a recharge water in lavas and gypsiferous deposits; otherwise a mixed

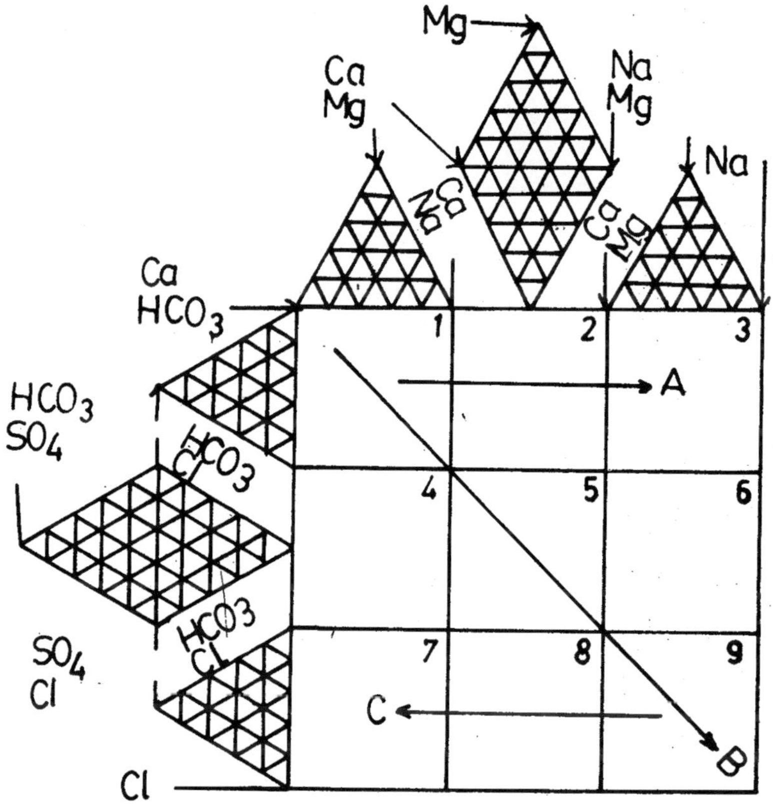

Fig. 12. Expanded Durov diagram

water or a water exhibiting simple dissolution may be indicated.

5. No dominant anion or cation, indicates waters exhibiting simple dissolution or mixing.

6. SO_4^{2-} dominant or anions indiscriminant and Na^+ dominant, is a water type not frequently encountered and indicates probable mixing influences.

7. Cl^- and Ca^{+2} dominant, is frequently encountered unless cement pollution is present in a well; otherwise the waters may result from reverse ion exchange of $Na^+ - Cl^-$ waters.

8. Cl^- dominant and no dominant cation indicates that the groundwater may be related to reverse ion exchange of $Na^+ - Cl^-$ waters.

9. Cl^- and Na^+ dominant frequently indicates end-point waters.

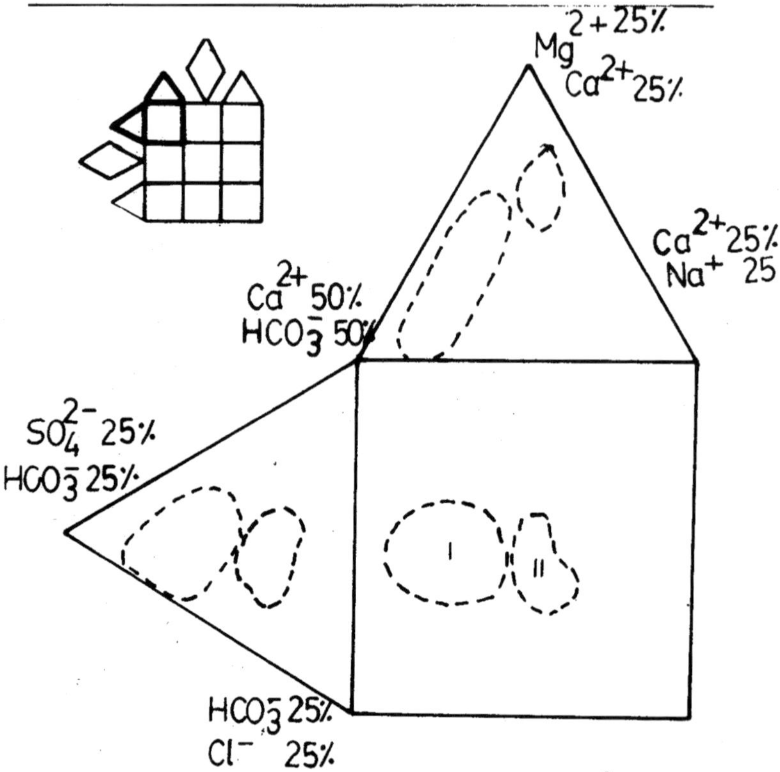

Fig. 13. Water type grouping in a sector of a Durov diagram 1. Ca^{2+} dominant recharge II. Mg^{2+} dominent recharge

The Durov diagram does not permit much distinction between $Na^+ - Cl^-$ waters.

Groundwaters can be classified or typed using the expanded Durov diagram. Durov diagram can also be used to represent minor ions for groundwaters classification (Fig. 14).

Zekai Sen and Al-Dakheel (1986) reported that neither trilinear nor the Durov diagram yields detailed information as given by the ion-concentration percentage frequency graphs.

Ion-Concentration percentage - Frequency Relationships

These graphs are proposed as supplements for the conventional trilinear diagrams. The aforementioned shortcomings can be avoided

5

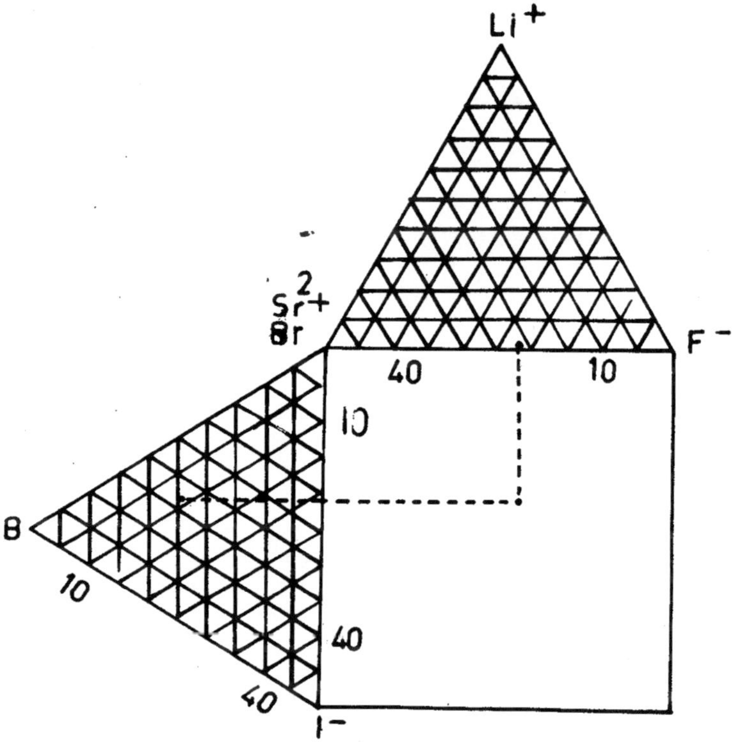

Fig. 14. Minor ion chemistry represented by a Durov diagram.

by these graphs (Fig. 15).

Any number of ions can be exhibited individually through these graphs (Fig. 15). In addition, it yields the frequency (number) of water samples falling within a certain percentage of epm values.

Equilibrium Studies

The theoretical background of the chemical equilibrium principles is widely available in various references among which are Klotz (1950), Garrels and Christ (1965), Krauskopf (1967), Stumm and Morgan (1970), and Back and Hanshaw (1970). The fundamental question is whether the water is in equilibrium with calcite or anhydrite. In order to simplify the calculations, Zekai Sen and AlDakheel (1986) have proposed practical graphs. (Figs. 16, 17 & 18).

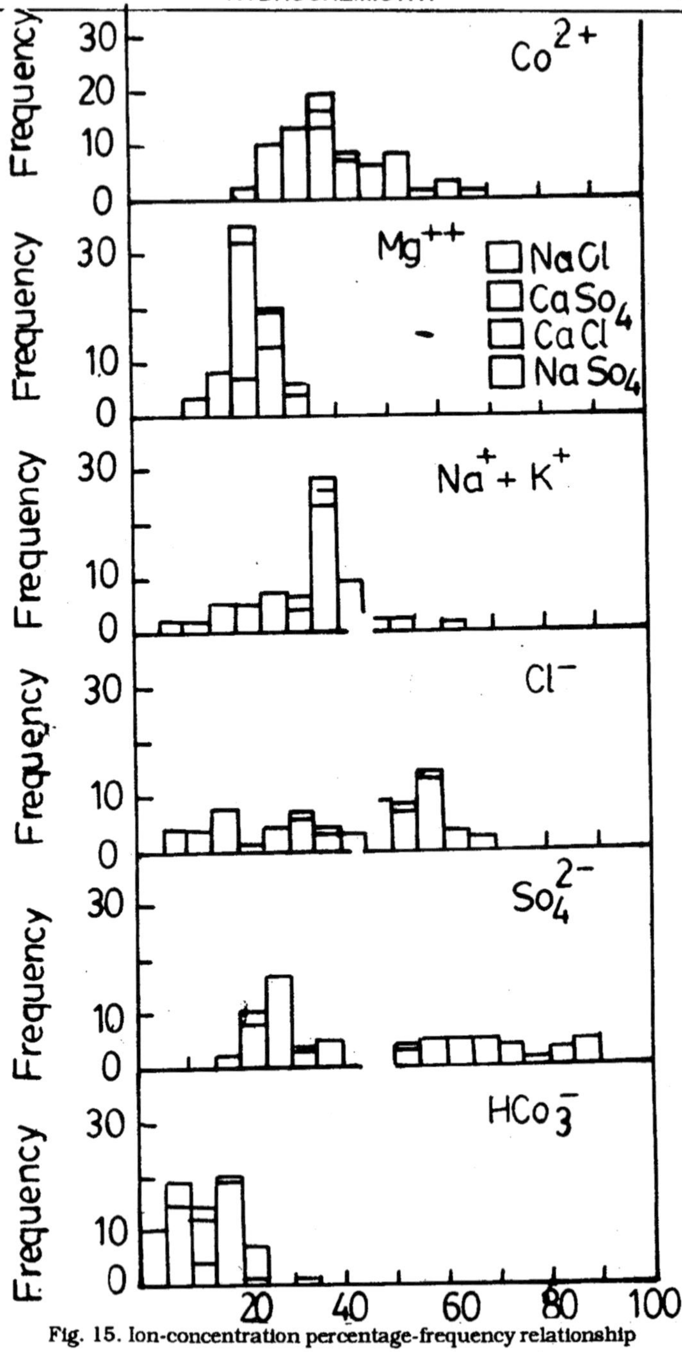

Fig. 15. Ion-concentration percentage-frequency relationship

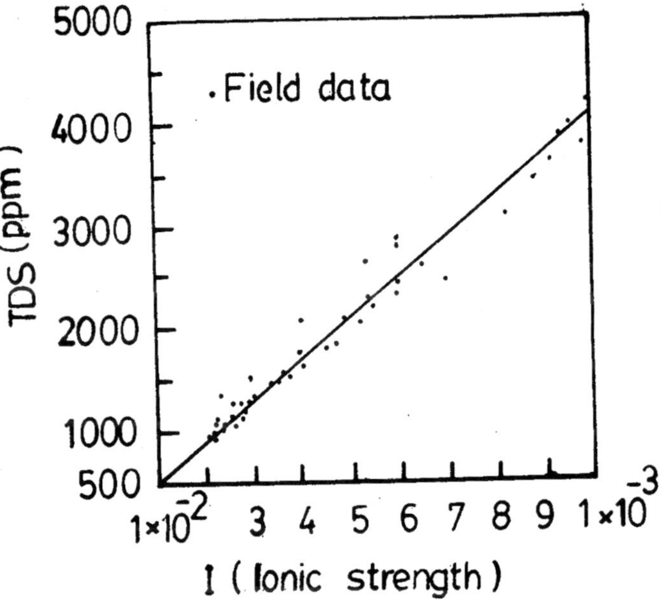

Fig. 16. TDS/Ionic strength relationsship

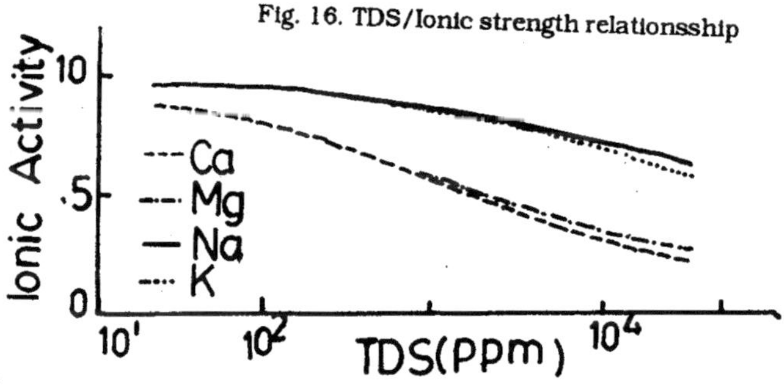

Fig. 17. Ion activity/TDS relation for cations.

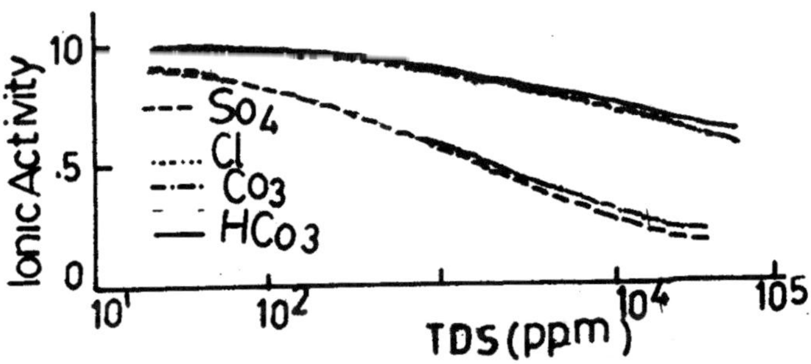

Fig. 18. Ionic activity/TDS relation for anions.

Ionic Strength/TDS/EC Relationships

If each ion in solution behaves independently, the solution is assumed as an ideal case. Otherwise, non ideal behaviours will appear as result of solute-solute and solute-solvent interactions. The magnitude of these interactions are dependent, among other factors, on the ionic concentrations and on ion electrical charge. The effects of these two major factors can be combined together through the ionic strength (IS) definition. IS can be determined as (Stumm and Morgan 1970)

$$IS = \frac{1}{2} \sum_{i=1}^{n} C_i Z_i^2$$

Where C_i is concentration of the i-th ion in moles/l; Z_i is the Oxidation number of the i-th ion; and n is the total number of ions. Although calculation of IS is possible after the laboratory analysis, its determination in the field through other composite variables will be very helpful for practical purposes. It is logical to expect a relationship between IS and TDS.

$$IS = 0.0000245 \times TDS$$

Substitution of TDS interms of EC into this last expression yields three relationships (Zekai Sen and Al-Dakheel, 1986).

$IS = 1.47 \times 10^{-5}$ EC (Nacl type and EC $\leq$ 2000)

$IS = 1.875 \times 10^{-5}$ EC (Mixed type)

$IS = 2.250 \times 10^{-5}$ EC (Ca SO_4 type and EC $\geq$ 2000)

Therefore, a careful measurement of EC in the field leads to approximate estimation of IS which avoids many lengthy calculations.

Activity Coefficient/ TDS Relationship

The Debye-Huckel (Stumm and Morgan, 1970) equation gives a relationship between the activity coefficients of different ions with the IS, provided some parameters in this equation are taken from tables for any specific ion. In order to avoid this long procedure, graphical relationship between activity coefficient and TDS are developed. For details see Zekai Sen and Al-Dakheel (1986).

Numerical approach to the classification of groundwater

This method is derived by Haddad (1981). All methods of classification depend on the variation of ionic ratio and ionic percentage. Generally the classification takes into consideration the relationship between major cations and anions. The ratios are arranged and given symbols as follows :

Cl/HCO_3; SO_4/HCO_3; Cl/SO_4; (Na+K)/Cl; Ca/(Na+K); Mg/(Na+K) and Ca/Mg.

The ratios being less than one, equal to one and more than one. A ratio of value less than one is given symbol 1, a ratio of value equal to one is given symbol 2, a ratio of value greater than one is given the symbol 3. This approach is applicable for providing rapid classification and evaluation of a large number of symbols (ElSchami and Rahim Hamdan, 1989).

Hydrochemical Modeling

During the movement, groundwater may dissolve, transport and deposit mineral matter. Therefore the major ion composition of groundwater reflects the overall effects of hydrology, mineralogy and weathering. Here, a review given by Narayanan Nair (1989) is included. A hydrogeochemical model is the integration of mathematical thermodynamic and kinetic relationships on which the geochemical calculations are based. These calculations are facilitated by the use of modern computers.

Hydrochemical modeling is largely a mental exercise but is facilitated by three types of calculations (1) speciation calculations (2) mass balance calculations and (3) calculations of mass transfer and reaction path.

Any development planning of groundwater resources needs a complete picture or model of the major ion characterisation of water. This could be obtained only through modeling the hydro-chemical faces (Balasubramanian et al., 1989).

The equilibrium speciation calculations determine the state of saturation of the water with respect of various minerals. These calculations can tell us whether a particular mineral would tend to dissolve on precipitate in a given aqueous environment.

In mass balance calculations, the amounts of mineral reactants and products which must dissolve and/or precipitate determined in order to account for the observed water quality. Mass balance calculations can also be used to solve problems of mixing of waters.

The mass transfer calculates the amounts of phases entering or leaving the aqueous phase required to account for the observed changes in composition between two water samples along a flow path. The mass transfer calculations are based on a thermodynamic model of the mineral-system and depends on the solubility of minerals, precipitation, starting water composition, assumed set of irreversible reactions, absorption and mixing of water. In general, mass balance relations are used to define the mass transfer during the chemical evolution of one water to another.

Numerical analysis of hydrochemical data have been attempted to determine the geochemical patterns of groundwater by Lawrence and Upchurch (1976, 1982) and Balasubramanian, (1986). Factor Analysis (FA), Principal Component Analysis (PCA) and Multiple Regression Analysis (MRA) are the widely used statistical or numerical concepts in parametric classifications and modeling studies (Balasubramanian, et. al. 1989). Detailed methods of these could be seen in Davis (1973).

Graphical representation of groundwater quality is limited because

(1) There is a finite number of variables that can be considered

(2) The variables are generally limited by convention to major ions, and

(3) Spurious relationships may be introduced by use of certain of the procedures.

Chayes (1960, 1971) has shown that by converting numbers to proportions with a constant sum, i.e., 100%, the numbers may no longer be independent. For example, in calculating the anion proportions for the trilinear diagram, chloride, sulphate and bicarbonate plus carbonate must some to 100% therefore, any increase in concentration in one or two of the variables forces an apparent, but perhaps false, decrease in the third variables. Clearly this response complicates interpretation of hydrochemical facies from trilinear diagrams. The Stiff diagrams, since they use raw rather than percent data, are not closed number systems. They depend upon pattern recognition, consider one sample at a time, and rely upon major ion distributions only (Matthew et. al., 1978).

To avoid the limitations of graphical interpretation of hydrochemical facies, Lawrence and Upchurch (1976) used factor analysis to classify water facies on the basis of chemical data. They showed that factor analysis can be used in much of the same way as the graphical methods with the added advantage of being able to map these water masses according to their factor classification.

The details of hydrogeochemical data taken from the Balasubramanian et al (1989) are listed below (Table-3). In FA and PCA, the dominant factors and principal components are extracted from the loadings matrix and the number of factors and components to be considered are determined based on an acceptable Eigen value, which normally is greater than 1.0. The scores (Factor scores in FA; Component Scores in PCA) scored by individual samples with reference to the extracted factors or components can give a clear picture about the dominant chemical ions of groundwaters over a basin. Models developed using the positive scores could be used for deterministic purposes.

In groundwater samples, EC (in Mmhos) is mainly dependent on the concentrations of major ions like Ca, Mg, Na+K, HCO_3, CO_3, Cl, NO_3 and SO_4. The functional relationship could be taken as EC = f(Ca, Mg, Na+K, HCO_3, CO_3, Cl, NO_3, SO_4) and the resulting regression coefficients of a data matrix could be used for the development of a model.

The dominant factors and principal components of groundwaters of the two river basins (Fig. 19) extracted from the loadings matrix are shown in table -3.

Table 3 : The dominant factors and principal componants of groundwater of The Tambraparni and Vaigai river basins (Balasubramanian et al., 1989)

Variables considered	Number of samples	Processing Analysis	River Basin
EC, pH, Ca, Mg, Na+k, HCO_3, CO_3, Cl, NO_3, SO_4	53	Factor Analysis	Tambraparni
-do-	-do-	Multiple Reg. Analysis	-do-
EC, Ca, Mg, Na+K, HCO_3 + CO_3, Cl, SO_4	67	Multiple Reg. Analysis	Vaigai
Ca, Mg, Na + K, HCO_3+CO_3, Cl, SO_4	-do-	Principal Component Analysis	-do-

Basin	Extracted Parameters	Chemical Parameters	Cumulative % of variance
Tambraparni	FACTOR I	Cl, EC. Ca and TDS	61.03
	FACTOR II	CO_3, pH, SO_4 and (Na+K)	75.47
	FACTOR III	HCO_3, (Na+k) and TDS	85.06
Vaigai	P.Component I	Cl, Mg, (Na+k)	52.50
	P.Component II	SO_4, HCO_3, Mg	72.75
	P.Component III	HCO_3, (Na+k)	87.69

Techniques in Basin wide Geochemical Interpretation

According to Deju (1971) based on the chemical analysis one can proceed to

(a) Calculate the salinities and alkalinities of the waters analyzed

(b) Correlate and differentiate waters on the basis of their geochemical index and

(c) Classify the waters according to the system of Chase Palmer

These three steps help in elucidating the origin of groundwater in a basin determine the chemical nature of aquifer rocks, and determine approximately the nature of the underground flow regime.

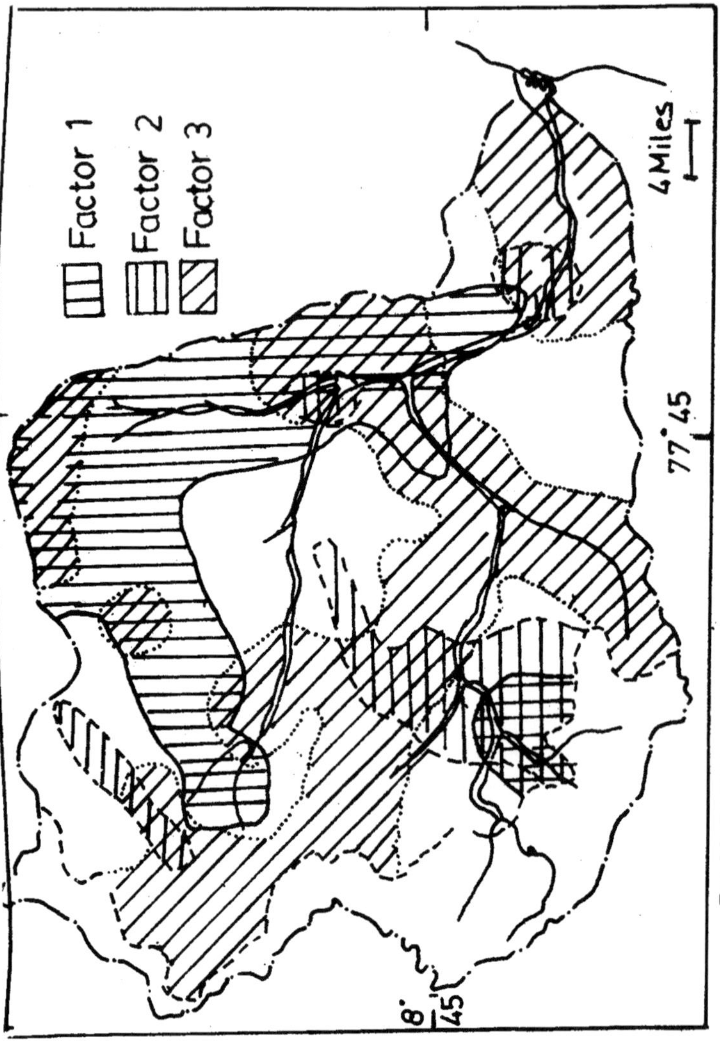

Fig. 19. Tambraparni basin hydrogeochemical model (by factor analysis). After Balasubramanian *et al.*, 1989.

In order to carry points a, b and c above one must begin by performing the following calculations (In epm.) :

Sum of Free Acids $= \Sigma SO_4 + Cl^- + NO_3 + NO_{2} = FA$

Sum of Week Acids $= \Sigma CO_3 + HCO_3 + PO_{4\text{-}4} = WA$

Sum of Alkalies $= \Sigma Na^+ + K^+ + Li^+ + NH_4 = A$

Sum of Earths $= \Sigma Ca^{++} + Mg^{++} = E$

On the basis of the four sums calculated above, one can classify the waters according to Palmer's system. This classification is as follows

Class I $FA < A$

Class II $FA = A$

Class III $FA > A$ and $FA < A+E$

Class IV $FA = A + E$

Class V $FA > A + E$

Once the class has been determined one can proceed to calculate the salinities and alkalinities. These quantities are defined as follows :

S1 : The primary salinity in a water is the amount of alkalies that are balanced by free acids, that is, the amount of salts resulting from the union of alkalies with free acids.

S2 : The secondary salinity is the amount of alkaline earths balanced by free acids.

S3 : The tertiary salinity is the excess of free acids over alkalies and alkaline earths. Waters where S3 > O have free acidity.

A1 : The primary alkalinity in a water is the amount of alkalies that are balanced by weak acids.

A2 : The secondary alkalinity is the amount of alkaline earths that are balanced by weak acids.

A3 : The tertiary alkalinity is that due to the presence of heavy metals. Most waters have A3 = O. If A3 > O some nearby mineralization must be present. The most common heavy metals causing tertiary alkalinity are copper, lead, zinc, iron and manganese. The letter M will denote the sum of heavy metals.

The values of the salinities and alkalinities vary depending on the class (Palmer) of the water. They can be calculated according to the following rules :

Class I	Class II
S1 = 2FA	S1 = 2FA
A1 = 2(A-FA)	A2 = 2E

A2 = 2E

A3 = 2M

A3 = 2M

Class III

S1 = 2A

S2 = 2 (FA-A)

A2 = 2 (A+E-FA)

A3 = 2M

Class IV

S1 = 2A

S2 = 2E

A3 = 2M

Class V

S1 = 2A

S2 = 2E

S3 = 2 (FA - A +E)

A3 = 2 (M - S3/2)

The salinities and alkalinities can be used to explain certain characteristics of the water. If the primary alkalinity is the predominant value, then the water must have percolated through sedimentary alkali beds. If the secondary alkalinity predominates, the water percolates through sedimentary strata rich in alkaline earths. Usually waters where A2 is predominant possess temporary hardness. When A3 predominates, it implies that the water passes through mineralized strata and possesses free acidity.

If the primary salinity is predominant, it implies that the water passes through volcanic terrain. If S2 predominates the water is usually one that has passed through sedimentary strata, volcanic in origin. Generally such waters have permanent hardness. Finally, if S3 predominates, the water must have percolated through mineralized terrain. Such waters usually possesses free acidity.

HYDROCHEMICAL CYCLE

The principal features of the geochemistry of natural waters can be shown by considering what may be called the 'hydro-chemical cycle'. The relations among the chemical character of water, the mineralogy of the environment and the circulation of water must be known in order to identify significant hydro-chemical problems.

The principal hydrochemical processes are;

(1) Wind blowing over the ocean carries sodium, chloride and other substances landward.

(2) As the water condenses, nitrogen, oxygen and carbondioxide of the atmosphere dissolve and are taken to the ground in rain and snow.

(3) Additional carbondioxide dissolves as water percolates

through soil rich in organic matter.

(4) Minerals dissolve and release many cations and anions

(5) Sulphide minerals oxidize to provide sulphate and some other constituents.

(6) Sulphate in solution is bacterially reduced and carbon-dioxide is formed.

(7) Minerals are precipitated as solubility products are exceeded and

(8) The water is returned to atmosphere by evaporation or transpiration, leaving behind the chemical products, or the water returns to the ocean as stream flow or groundwater discharge, carrying dissolved and perhaps suspended matter with it. This is a simplified description of the hydrochemical cycle and covers only the most important processes (Back and Hanshaw, 1965).

Chemical processes reflect the hydrochemical facies in the lighologic environment and the prevailing groundwater flow patterns. The mineralogic framework determines which facies will develop and the vertical and horizontal flow pattern controls the distribution of the facies. The groundwater flow pattern is controlled primarily by the distribution of head within the hydrologic system. The distribution of head is controlled by the relative elevations and locations of recharge and discharge areas and by variations in the transmissibility of the system. Therefore, the three fundamental factors controlling the chemical character of groundwater are the mineralogy, transmissibility and regional topography of the hydrologic system. Studies that omit considerations of any one of these controls cannot satisfactorily explain the genesis of the chemical character of groundwater (Back and Hanshaw, 1965).

Geochemical Computer Codes

Computer models offer an incredibly powerful tool for understanding the chemical state of natural waters and for predicting the behaviour of such waters under a variety of hypothetical conditions. Atleast 50 computer codes are available in literature ranging from those that can be run on a programmable hand-calculator to massive programs that require a super computer. Some of the best-known of the comprehensive models of natural waters are PATHI (Helgeson et al, 1970), MINEQL (Morel and Morgan, 1972), WATEQ (Truesdell and Jones, 1973), WATEQF (Plummer et al, 1976) EQ 3/6 and EQ3NR (Wolery, 1979, 1983), GEOCHEM (Sposito and Mattigod, 1980), PHREEQE (Parkhurst et al, 1980). Nordstorm and Ball (1984) and Sposito (1985) deal with the descriptions of the mathematical and chemical principles involved in the models. Briefly, the codes generally solve a set of simultaneous non-linear equation that describe mass-balance, electrical neutrality and chemical equilibria. It is a puzzling question as to why so many geochemical models have been written. Many of them are duplicative and there is no apparent reason for having such a large number of codes (Runnels, 1987).

Major Types of Geochemical Models

There are three major types of computer models in aqueous geochemistry (Nordstrom and Ball, 1984). They are (1) equilibrium models (2) mass-transfer models (3) mass-transport models.

Equilibrium models solve the equations those describe equilibrium reactions and mass balances on the dissolved elements. Ex : WATEQ, WATEQF, GEOCHEM, EQ3NR.

Mass-transfer models, in addition to the equilibrium models, are able to compute new compositions of the water as selected minerals and compounds are precipated or dissolved. Ex : PATH1, EQ6 MINTEQ.

Mass-transport models combine chemical reactions with the movement of fluid. Ex : RIVEQL, CHEMTRN, TRANQL.

There are various limitations, especially in the thermodynamic data base used by the various models. The results obtained can be used to establish base line geochemical conditions for potential repository sites and to simulate the geochemical behaviour of transported radii nuclids.

REFERENCES

Back W., (1960) Origin of hydrochemical facies of groundwater in the Atlantic coastal plain. Internal. Geol. Cong., 21 Session, Pt.1, Geochemical cycles, PP. 87-95.

Back W., (1966) Hydrochemical facies and groundwater flow patterns in northern part of Atlantic Coastal Plain. U.S. Geol. Surv. Porf. Paper, 498A, 42 PP.

Back W and Hanshaw B. B., (1965) Chemical geohydrology in V.T. Chow (Ed.)., Advances in Hydroscience, Academic Press, Inc., New York. V. 2., PP. 49-109.

Back W and Hanshaw B. B., (1970) Comparison of chemical- hydrogeological of the carbonate peninsulas of Florida and Yucatan, Jour. Hydrol., 10, PP. 330-368.

Balasubramanian A Tirugnana Sambandham R Chellasamy R and Sastri JCV., (1989) Digital modeling of the Hydrochemical facies of groundwaters in Tamailnadu, India. In International workshop on Appropriate methodologies for development and management of groundwater resources in developing countries, NGRI, Hyderabad., V. II., PP. 761-767.

Balasubramanian A., (1986) Hydrogeological investigations in the Tambraparni river basin, Tamil Nadu. Unpublished Ph.D. Thesis, Univ. of Mysore, Mysore, 349 PP.

Burdon DJ and Mazloum, S., (1958) Some chemical types of groundwater from Syria. UNESCO symp. Tehran, UNESCO, Paris, PP. 73-90.

Chayes F., (1960) On correlation between variables of constant sum. Jour

Geophys. Res. 65 PP. 4185-4193.

Chayes. F., (1971) Ratio Correlation, Univ. of Chicago Press Chicago, 99 PP.

Davis John C., (1973) Statistics and data analysis in Geology. John Wiley & Sons, New York, 550 PP.

Deju R.A., (1971) Regional hydrology fundamentals, Gordon and Breach Science Publishers, New York, PP. 176-189.

Durov S.A., (1948) Natural Waters and Graphic representation of their composition. Dokl. Akad. Nauk. SSSR, 59, PP. 87-90.

EIS Chami F and Rahim Hamdan A., (1989) Hydrochemical classification of the groundwater of the Wadi Albih well field in the northern part of the United Arab Emirates in Intern. Symp. NGRI, Hyderabad, PP. 853-866.

Fetter C.W., (1988) Applied hydrogeology, Merril Publishing Co., USA 592 PP.

Garrels RM and Christ CL., (1965) Solutions, minerals and equilibria. Harper and Row, New York, 497 PP.

Haddad R.H., (1981) A proposed new approach to the classification of ground water. First symposium on the future development of water resources in the Gulf and Arabian Peninsula, Kuwait Univ. PP. 47-61.

Helgeson H. C., Brown T. H, Nilgrini A. and Jones T. A.., (1970) Calculation of mass transfer in geochemical processes involving aqueous solutions. Geochemica. et. Cosmochemica. Acta, 34., PP. 569-592.

Hem J.D., (1959) Study and interpretation of the chemical characteristics of natural water. U.S. Geol, Surv. Water Supply Paper, 1473, 363 PP.

Hill R. A., (1940) Geochemical patterns in the Cochlea Valley California. Trans. Am. Geophys. Union, 21, PP. 46-49.

Klotz I. M., (1950) Chemical thermodynamics, Printice Hall, New Jersy.

Krauskoph K., (1967) Introduction to Geochemistry, Mc Graw - Hill, Inc., New York, 721 PP.

Lawrence F. W. and Uppchurch SB., (1976) Identification of geochemical patterns in groundwater by numerical analysis. Proc. 12th Ann. Symposium, advances in groundwater hydrology, Am. water resources Assoc. PP. 199-214.

Lewrence F. W. and Upchurch SB., (1982) Identification of recharge areas using geochemical factor analysis, Groundwater, 20 (6), PP. 680 -687.

Lloyd J. W., (1965) The hydrochemistry of The aquifers of north - eastern Jordan, Jour. Hydrol. 3, PP. 319-330.

Lloyd J. W., and Heathcote JA., (1985) Natural inorganic hydrochemistry in relation to ground water - An Introduction. Clarendon Press, Oxfrod, 296 PP.

Matthew G. Dalton and Sam B. Upchurch (1978) Interpretation of hydrochemical facies by factor analysis. Groundwater, 16(4).

Morel F and Morgan J. J., (1972) A numerical method for computing equilibria in aqueous systems. Envi.Sci. and Tech. 6, PP. 58 - 67.

Narayanan Nair V. (1989) Hydrogeochemical modeling, International symp. NGRI, Hyderabad, V II, PP. 903 - 906.

Nordstrom D. K. and Ball J.W., (1984) Chemical models, computer programs and complexation in natural waters. In C.J.M. Kramerand J.C. Duinker (eds), Complexation of trace metals in natural waters. Martinus Nijhoff/Dr.

W. Junk Publ. The Netherlands, PP. 149 - 164.

Palmer C., (1911) The geochemical interpretation of water analysis. U.S. Geol. Surv. Bull. 479, 31 PP.

Parkhurst D. L. Thorstenson D. C. and Plummer L. N., (1980) PHREEQE - A computer program for geochemical calculations. U.S. Geol. Surv. Water resources Investigation, 80 - 96, 210 PP.

Piper, A. M., (1944) A graphic procedure in the geochemical interpretation of water analysis. Am. Geophys. Union.trans. 25, PP. 914 - 923.

Piper A. M., (1953) A graphic procedure in the geochemical interpretation of water analysis. U.S. Geol Surv. Groundwater Note. 12.

Plummer L. N. Jones BF and Trusidell AH., (1976) WATEQF - a fortran IV version of WATEQ, a computer for calculating chemical equilibrium of natural waters. U.S. Geol. Surv. Water Resources Investigation 76 - 13, 61 PP.

Runnels D Donald (1987) American research in the geochemical modeling of groundwater. Jour. Geol. Soc. India. 29, PP. 135-144.

Schoellar H., (1962) Les eaux souterraines Massio, Paris.

Sposito G., (1985) Chemical models of inorganic pollutants in soils. CRC critical reviews in environmental control, 15(1), PP. 1-24.

Sposito G. and Mattigod S.V., (1980) GEOCHEM : A computer program for the calculation of chemical equilibria in soil solutions and other natural water systems. Dept. Soil and environmental Sciences Report, Univ. California, Riverside California, USA, 92 PP.

Stiff H.A., (1951) The interpretation of chemical water analysis by means of patterns : J. petrol. Tech, 3, PP. 15-17.

Stumm W and Morgan J.J., (1970) Aquatic Chemistry. John Wiley & Sons. New York, 583 PP.

Tuesedell A.H. and Jones B.F., (1973) WATEQ, a computer program for calculating chemical equilibria of natural waters U.S. Geol. Surv. Jour. Res. 2, PP. 233-274

Winograd I.J. and Farlekas G.M., (1974) Problems in 14_C (dating of waters from aquifers of deltaic origin. In Isotope Hydrology, International Atomic energy Agency, Vienna, PP. 69-93.

Wolery T.J., (1979) Calculation of chemical equilibria between aqueous solution and minerals : the EQ 3/6 software package : Lawrence Livermore National Lab., UCRL-52658, Livermore, california, U.S.A.

Wolery T.J. (1983) EQ_3NR, a computer program for geochemical aqueous speciation-solubility calculations : Users guide and documentation : Lawrence, Livermore National Lab, UCRL - 53414, Livermore, California U.S.A. 191 PP.

Zakai Sen and Al Dakheel A., (1986) Hydro-chemical facies evaluation in Umm Er Radhuma Limestone, Eastern Saudi Arabia, Groundwater, 24 (5), PP. 626-635.

5

WATER QUALITY

Dr. P. SANKARA PITCHAIAH

Dept. of Geology, Nagarjuna University,
Nagarjuna Nagar, 522 510 (A.P.)

The term 'Water Quality' is an extremely confusing concept meaning different things to different people. Basically the human mind tends to perceive water quality as being good if desirable water uses are possible, and bad if water quality involves water use problems. The complex composition of natural water further complicates this simple idea. The content of one particular substance may be well below any acceptable concentration limit for one type of water use but may be prohibitive for another. There is an increasing consensus now that the term water quality refers to the 'sustainability of water for its desired use'. This means that there is no universally acceptable set of concentrations, as different water uses have different quality criteria to sustain desired water uses (Falkenmark and Allard, 1991).

In groundwater resource evaluation, the quality of groundwater is of nearly equal importance to quantity. The physical, chemical, biological and radiological characteristics of groundwater determine its usefulness for domestic, industrial and agricultural water supplies. The groundwater quality data give important clues to the geologic history of rocks and indications of groundwater recharge, discharge, movement and storage.

A knowledge of water quality is also essential for rational management of water and land resources (Karanth, 1989).

Water Quality and Hydrologic Cycle

Precipitation contains dissolved substances, largely determined by the air quality and wind patterns of the region. In areas where there is

GROUND WATER
Editor: P. Sankara Pitchaiah, © 1994 Scientific Publishers, Jodhpur

heavy air pollution the atmosphere is a complex chemical system controlled by dozens of chemical and photochemical transformations (Mac Cracken and Sauter, 1975), Atmospheric water in such areas accumulates carbondioxide, nitrates and inorganic forms of phosphorus and sulphur. This moisture becomes chemical laden precipitation hundreds of miles from the original pollution source (USEPA, 1979).

Data from areas relatively unaffected by human activity indicate that much of the instream burden of nitrogen comes from precipitation, while only a small portion of the instream burden of phosphorus is precipitation borne (Stewart and Markello, 1974).

The majority of chemicals in precipitation are absorbed by soil and become part of the nonpoint-source pollution which may eventually be washed into streams or lakes. Precipitation which falls directly on surface waters can have varied effects. Dilution by relatively clean rainfall can improve water quality in lakes. Precipitation containing large amounts of nitrogen and sulphur is highly acidic and can disrupt the entire biochemical environment.

Surface runoff

As the water moves across the land surface during or after a storm, it transports dissolved and suspended materials which have been picked up along the path of flow. Wash off materials of most concern include sediment, mineral salts, heavy metals, nutrients, pesticides, biodegradable organics, and microbial pollution. Dungion and Crawford (1976) summarizes the water quality charactiristics of runoff with various land uses.

Infiltration

To a large extent soil type and texture determine the interactions in which water and its chemical burden are involved. Fine textured soils with high clay content do not drain well and retain large amounts of water for long period. Aeration in these soils is limited and processes such as organic decomposition, ammonia volatilization, and nitrification are retarded. Although clay particles are active sites for ion exchange and absorption, the effective absorption of dissolved materials is reduced since the water molecules are polar and compete for adsorption sites. Coarse-textured soils conduct large quantities of air and water and oxidative processes are encouraged. At the same time the rates of evaporation, lateral transmissibility, and percolation are higher for coarse textured soils.

As a rule, passage through the soil profile results in purification of water because of adsorption, volatilization, decomposition or degradation, nitrification, de-nitrification and plant uptake (Linsley et al., 1988).

Water Quality and Rock Types

Intrusive Igneous and Metamorphic Rocks

Intrusive igneous and metamorphic rocks have very little primary porosity. Secondary porosity is the main promoter of the groundwater storage in these rocks which develops due to tectonic movements, erosion of over burden, rock loading and unloading of glaciation, shrinking during cooling of the rock mass and the compression and tensional forces caused by regional tectonic stresses.

In general, the amount of fracturing in crystalline rocks decreases with depth (Davis and Turk, 1964). However, it is not always true (Coats et al., 1983; Haimson and Doe, 1983; Daniels et al., 1983; Coutre et al., 1983).

Chemical weathering of crystalline rock can produce a weathering product called saprolite. This material has porosities of 40 to 50 per cent and a specific yield of 15 to 30 per cent. It acts as a reservoir, storing infiltered water and releasing it to wells interesecting fractures in the underlying crystalline rock (Welby, 1984).

Though, the minerals present in crystalline rocks (quartz, feldspars, amphiboles, and mica) are relatively insoluble, it is not absolute. The rate of solution is slow, and the dissolved matter in solution at any time is small. The action of meteoric waters on the minerals present involves, first solution of the mineral and then hydrolysis. Because the carbondioxide universally present in them, and the action of the water on the minerals is accelerated, the amount of acceleration depending on the amount of carbondioxide in solution. Even in the presence of carbonic acid, the silicate minerals that make up the crystalline rocks are relatively resistant to attack Liegeois (1928) immersed quartz and compact granite, syenite and diabase in carbonated water and found no alteration on microscopic examination even after six months, except one granite, in which feldspar was attacked.

Water from crystalline rocks is therefore usually low in dissolved mineral matter because of the slow weathering of silicates. Arid regions are an exception, here evaporation can lead to high concentrations of dissolved solids.

High silica content is characteristic of groundwater in crystalline rocks (Hem 1970). The alkalis and alkaline earths are liberated by weathering of silicates ; however their ratios in groundwater are generally different from those in the original rocks.

The iron liberated by weathering goes into solution, although some of it is precipitated as amorphous iron hydroxides. Sulphur and chloride ions are present in low concentrations.

Groundwater in Volcanic Rocks

Volcanic rocks cool and solidify quickly which gives then radically different porosity characteristics. Volcanic rocks include lava flows : deposits of ash and cinders occurs in loose unconsolidated piles. Pyroclastic deposits have higher porosities.

Hydraulic conductivity of volcanic rocks such as lava flows and under beds is typically quite high and is low for ash beds, intrusive dikes, and sills. The groundwater contain more silica. Alkaline earths are dominant over alkalis, except in dry regions.

Groundwater in Sedimentary Rocks

The three main types of sedimentary rocks, in the order of their estimated abundance are shale (82%) sandstone (12%) and limestone (including chalk and dolomite) 6%.

Sandstone show in addition to the joints and permeability a variable matrix permeability. The effective pore spaces present offer extensive contact surface and cause relatively low permeability.

The character of the dissolved solids depends on the material in the sandstone. Pure siliceous material without cement leads to groundwater with low total dissolved solids. Groundwater in sandstones with soluble cementing materials may show appreciable amounts of sulphate, chloride, sodium, magnesium and calcium ions.

Shales have higher porosities, however it decreases significantly as a consequence of the mechanical pressure forces. In general the greatest decrease in porosity occurs because of onset of diagonotic changes approximately, as far down as 500m. In the zone between 1000 and 3500m the porosities decrease proportionally to the depth. At 4000m depth porosities below 1% have been measured (Harvey, 1982).

Since water movement is very slow in these rocks due to poor permeability, there is a long period of rock-water contact. Groundwater often rich in dissolved solids. Sulphates and chlorides are high and associated with calcium and magnesium and sodium respectively. Groundwater in argillaceous rocks has comparatively higher iron and fluoride content and relatively low pH.

The primary porosity and permeability of the limestones can be variable, ranging from practically impermeable dense pore-free rocks to extremely permeable sediments rich in voids.

Groundwater in limestones contains more alkaline earth and carbonate ions, relative to alkalis, chloride and sulphide. The low solubility of carbonates and the limited contact surface explain the small content of dissolved solids. The pH of the groundwater from carbonate rocks is generally above 7.

For dense limestone the greater part of the residue after evaporation consists of HCO_3^- and Ca^{2+} ions. SO_4^{2-} and Cl^- are present only in very small amounts. The water from fine grained limestones shows slight increase in the SO_4^{2-} and Cl^- contents, where as from porous limestones has increased HCO_3^- and Ca^{2+} content.

Water Quality Parameters

Water quality parameters can be divided into four categories, i.e.,(1) Physical (2) Chemical (3) Biological (4) Radiological and are presented in table - 1a. Health related inorganic constituents are shown in table - 1b. Indian standard specifications for drinking water are shown in table - 1c.

Table 1a. Some Important Water Quality Parameters

Physical	Chemical	Biological	Radiological
Colour	Acidity	Virus	Uranium
	Hardness	Bacteria	
Foam and Froth	Alkalinity	Algae	
Odour	Ammonia	Other Nuisance-	
	Nitrates	organisms	
Temperature	BOD		
Turbidity	Calcium		
	Magnesium		
	COD		
	Chlorides		
	Conductivity		
	DO		
	Iron		
	pH		
	Sulphates		
	Phosphates		
	Radox potential		
	Sodium		
	Potassium		

Table 1b. Health Related Inorganic Constituents

Arsenic	Lead
Asbestos	Mercury
Barium	Nickel
Cadmium	Nitrate
Chromium	Selenium
Cynide	Silver
Fluoride	Sodium
Hardness	Zinc
Iron	

Table-1c : Indian standard specifications for drinking water.
IS 10500-1983

S. No.	Substance or characteristics	Requirement desirable limit	undesirable effects outside the desirable limit	Desirable Essential	Remarks
(i)	Colour, Hazen Units, Max.	10	Above 10, consumer acceptance decreases	Essential	May be extended to 50 only if toxic substances are not suspected in absence of alternate sources
(ii)	Odour	Unobjectionable	—	Essential	(a) Test cold and when heated (b) Test at several dilutions
(iii)	Taste	Agreeable	—	Essential	(a)Test to be conducted only after safety has been established.
(iv)	Turbidity NTU Max.	10	Above 10, consumer acceptance decrease	Essential	May be extended up to 25, in absence of alternate sources.
(v)	pH value	6.5 to 8.5	Beyond this range the water will affect the mucous membrane and/or water supply system.	Essential	May be relaxed up to 9.2 in absence of alternate sources.
(vi)	Total hardness (as ($CaCO_3$) mg/L Max.	300	Encrustation in water supply structure and adverse effects on domestic use	Essential	May be extended up to 600, in the absence of other sources
(vii)	Calcium (as Ca) mg/L, Max.	75	Encrustation in water supply structure and adverse effects domestic use.	Desirable	May be extended up to 200, in the absence of other sources.

1.	2	3	4	5	6
(viii)	Magnesium (as Mg) mg/L, Max.	30	Encrustation in water supply structure and adverse effects domestic use.	Desirable	May be extended up to 100, in the absence of other sources.
(ix)	Copper (as Cu) mg/L Max.	0.05	Astringent taste discoloration and corrosion of pipes fittings and utensils will be caused beyond this.	Desirable	May be relaxed up to 1.5
(x)	Iron (as Fe) mg/L, Max.	0.3	Beyond this limit taste/appearance are affected has adverse effect on domestic uses and water structures, and promotes iron bacteria	Essential	May be extended up to 1.0 in absence of alternate sources.
(xi)	Manganese (as Mn) mg/L, Max.	0.1	Beyond, this limit taste/appearance are affected on domestic uses and water supply structure.	Desirable	May be extended up to 0.5 where alternate source is not available.
(xii)	Chlorides (as Cl) mg/L, Max.	250	Beyond this limit, taste corrosion and palatibility are affected.	Essential	May be extended up to 1000 in the absence of other alternate sources.
(xiii)	Sulphate (as SO_4) mg/L, Max.	150	Beyond this causes gastro-intestinal irritation when magnesium of sodium are present	Desirable	May be extended up to 400 provided magnesium (as Mg) does not exceed 30.

1.	2	3	4	5	6
(xiv)	Nitrate (as NO_3) mg/L, Max.	45	Beyond this methane moglobinemia takes place	Derlable	No relaxation
(xv)	Fluoride (as F) mg/L,	0.6 to 1.2	Low fluoride levels are linked with dental caries above 1.5 it may cause fluorosis.	Desirable	If the limit is below 0.6 water source should be rejected but suitable public health measures should be taken. Maximum limit may be extended to1.5 if no better alternate source is available.
(xvi)	Phenolic compounds (as C_6H_5HO) mg/L Max.	0.001	Beyond this it may cause objectionabe taste and odour	Desirable	May be relaxed up to 0.002
(xvii)	Mercury (as Hg) mg/L, Max.	0.001	Beyond this the water becomes toxic.	Desirable	No relaxation of this limit is allowed. To be tested when pollution is suspected.
(xviii)	Cadmium (as Cd) mg/L, Max.	0.01	Beyond this the water becomes toxic.	Desirable	No relaxation of this limit is allowed. To be tested when pollution is suspected.
(xix)	Selenium (as Se) mg/L, Max.	0.01	Beyond this the water becomes toxic.	Desirable	No relaxation of this limit is allowed. To be tested when pollution is suspected.
(xx)	Arsenic (as As) mg/L, Max.	0.05	Beyond this the water becomes toxic.	Desirable	No relaxation of this limit is allowed. To be tested when pollution is suspected.
(xxi)	Cyanide (as Cn) mg/L, Max.	0.05	Beyond this limit the water becomes toxic.	Desirable	No relaxation. To be tested when pollution is suspected.
(xxii)	Lead (as Pb) mg/L, Max.	0.1	Beyond this limit the water becomes toxic.	Desirable	No relaxation being a health parameter. To be tested when polution/plumbosolvency is suspected.

1.	2	3	4	5	6
(xxiii)	Zinc (as Zn) mg/L, Max.	5.0	Beyond this it can cause astringent taste and an opalescence in water	Desirable	May be relaxed up to 10.0. To be tested when pollution is suspected.
(xxiv)	Anionic detergents (as MBAS) mg/L, Max.	0.2	Beyond this limit it can cause a light froth in water	Desirable	May be relaxed up to 1.0. To be tested when pollution is suspected.
(xxv)	Chromium (as Cr^{6+}) mg/L Max.	0.05	May be carcinogenic above this limit	Desirable	No relaxation. To be tested when pollution is suspected.
(xxvi)	Polynuclear aromatic hydrocarbons (as PAH) mg/L, Max.	-	May be carcinogenic	Desirable	-
(xxvii)	Mineral Oil mg/L, Max.	0.01	Beyond this limit undesirable taste and odour after chlorination takes place.	Desirable	May be relaxed up to 0.03. To be tested when pollution is suspected.
(xxviii)	Residual free chlorine mg/L, Max	0.2	-	Essential	To be applicable only when water is chlorinated. Tested at consumer end. When protection against viral infection is required, it should be Min. 0.5 mg/L.
(xxix)	Pesticides	Absent	toxic	Desirable	-
(xxx)	Radioactive	-	-	Desirable	-
	(a) Alpha emitters μc/ml Max.	10^{-8}			
	(b) Beta emitters μc/ml Max.	10^{-7}			

Physical Parameters

Colour

Colour in groundwater mainly results from degradation process in the natural environment. It may occur due to the presence of humic acids, fluvic acids, metallic ions such as iron and manganese, suspended matter, industrial wastes, etc.

Colour is expressed in terms of Hazan standard unit which defined as the colour produced by 1 mg/1 of platinum in the form of chloroplatinic acid in the presence of 2 mg/1 of cobaltous chloride hexahydrate.

Foam

It is a suspension or dispersion of gases or air bubbles in a liquid medium. Foams are caused by the presence of foaming substances such as synthetic detergent, soaps, protein, etc. Sometimes, it traps pathogenic organisms.

Odour and Tastes

The odours and tastes are present mainly due to dissolved impurities often organic in nature.

The odour may be of natural origin, caused by living and decaying aquatic organisms and accumulation of gases like ammonia and hydrogen sulphide, etc. Many algae also impart taste and odours to water. Sometimes reagents added to water supply systems may also produce odour and tastes.

Many inorganic chemicals are also capable of imparting a characteristic taste (Table 2).

Table 2. Some Inorganic chemicals with their threshold limits imparting tastes to the water (Voznaya, 1981)

Salt	Concentration (mg/l) Indefinite, slight hardly distinguishable taste	Repulsive taste
NaCl	150	150 salty
$MgCl_2$	100	400 bitter
$MgSO_4$	200	500 bitter
$CaSO_4$	70	150 astringent
KCl	350	700 bitter
$FeSO_4$	1.5	5 chalydeate
$MnCl_2$	2	4 stagnant
$FeCl_2$	0.3	0.5 stagnant

The objectionable tastes and odours are sometimes rejected on the ground of aesthetic value. Some organic substance imparting taste and odours may also be toxic. The tastes and odours in the water are also not suitable in food, pharmaceuticals and beverage industries.

Temperature

Variations in solar energy received at the earth's surface create periodicities, both diurnal and annual in temperature below ground surface. In marked contrast to the large seasonal variation of surface water temperature groundwater temperature tend to remain relatively constant.

Below the zone of surface influence groundwater temperatures increases approximately 2.9°C for each 100 m of depth in accordance with the geothermal gradient of the earth's crust.

Temperature must be measured immediately after collecting the sample. It is basically important for its effect on the chemistry and biological reactions in the organisms in water. A rise in temperature of the water leads to the speeding up of the solubility of gases and amplifies the taste and odours. Water in the temperature range of 7°C to 11°C has pleasant taste and is refreshing. At higher temperatures with less dissolved gases, the water becomes tasteless.

Temperature is also important in the determination of various parameters such as pH, conductivity, saturation level of gases and alkalinity. Data on temperature is also required by the industries in heat transmission calculation, cooling towers and process use.

Turbidity

Turbidity in water is caused by the substances like clay, silt, organic matter, phytoptankton and other microscopic organisms. It is actually the expression of optical property.

Turbidity makes the water unfit for the domestic purposes and some industries, however it is negligible in groundwater. Sometimes turbidity is measured to evaluate the performance of water treatment plants.

Chemical Parameters (pH)

pH is the measure of the intensity of acidity or alkalinity and measures the concentration of hydrogen ion in water.

Most groundwater has pH values between 5.8 and 8.0. Very high pH values, above 8.5 are usually associated with sodium carbonate bicarbonate waters, moderately high pH values are comonly associated with waters high in bicarbonates, very low pH values may be associated with small amounts of mineral acids from sulphide sources or with organic acids.

pH has no direct adverse affect on health, however, a low value below 4 will produce sour taste, and higher value above 8.5 an alkaline taste.

Electrical Conductivity (EC) and Total dissolved solids (TDS)

Conductivity is the measure of capacity of a substance or solution to conduct electric current. It is reciprocal of resistance. Conductivity if measured of a cube with each side of 1 cm at 25°C is called 'specific conductance.' As the most of the salts in the water are present in the ionic forms and capable of conducting current, conductivity is a good and rapid measure of the total dissolved solids. As the dissolved solids in the highly mineralized waters are usually more than 65% conductivity, the value of dissolved solids, as a general rule, can be obtained by multiplying it by a factor of 0.65. However, it is difficult to get accurate values by this calculation, as capacity to conduct the current depends on number and kinds of ions present, their relative charge, and freedom of ions to act as conductors. Solids should not be estimated from the conductivity value, if it exceeds 50 mmho (50,000 μ mho).

Electrical conductivity is expressed as mhos per centimeter. It is such a large unit that most natural water has a value much less than one unit. For convenience these values can be reported as milli mhos ($EC \times 10^3$) or micro mhos ($EC \times 10^6$) at 25°C ($EC = 0.00255$ mhos/cm $= EC \times 10^3 = 2.55$ milli mhos/cm $= EC \times 10^6 = 2550$ micro mho/cm.)

Total dissolved solids denote mainly the various kinds of minerals present in the water. In natural waters, dissolved solids are composed mainly of carbonates, bicarbonates, chlorides, sulphates, phosphates and nitrates of calcium, magnesium, sodium, potassium, iron and manganese, etc.

Concentration of dissolved solids is an important parameter in drinking water (Table - 3) and other water quality standards. They give a particular taste to the water at higher concentration and also reduce its palatability.

Table 3. Concentration of TDS and drinking water

Excellent	:	Less than 300 mg/l
Good	:	between 300 and 600 mg/l
Fair	:	between 600 and 900 mg/l
Poor	:	between 900 and 1200 mg/l
Unacceptable	:	greater than 1200 mg/litre

Source : Bruvold et. al., (1967)

Dissolved Oxygen (DO)

Dissolved oxygen is one of the most important parameter in water quality assessment and reflects the physical and biological processes prevailing in the water. Its presence is essential to maintain the higher forms of biological life in the water.

Low oxygen in water can kill fish and other organisms present in water. Low oxygen concentrations are generally associated with heavy contamination by organic matters and the water characterizes with insipid taste.

Biological Oxygen Demand (BOD)

Biological Oxygen Demand is the amount of oxygen utilized by microorganisms in stabilizing the organic matter. BOD approximates the amount of oxidizable organic matter present in the solution. The BOD values are very useful in process design and loading calculations as well as the measure of treatment plant efficiency and operation.

Chemical Oxygen Demand (COD)

Chemical Oxygen Demand is the oxygen required by the organic substances in water to oxidize them by a strong chemical oxidant.

The COD test is very important parameter in management and design of the treatment plant because of its rapidity in determination.

Alkalinity

The alkalinity of a water is a measure of its capacity to neutralize acids. The alkalinity of natural waters is due to the salts of carbonates, bicarbonates, borates, silicates and phosphates along with the hydroxyl ions in the free state. However, the major portion of the alkalinity in natural water is caused by hydroxide, carbonate and bicarbonates which may be ranked in order of their association with high pH values. Alkalinity values provide guidance in applying proper doses of chemicals in water and waste water treatment processes, particularly in coagulation softening and operational control of anaerobic digestion. The ratio of alkalinity to that of alkaline earth metals is a good parameter determining the suitability of irrigation waters.

Acidity

Acidity of the water is its capacity to neutralize a strong base and is mostly due to the presence of strong mineral acids, weak acids (carbonic acetic acids) and the salts of strong acids and weak bases (e.g., ferrous suflphate, aluminum sulphate, etc.). These salts on hydrolysis produce a strong acid and metal hydroxides which are sparingly soluble thus producing the acidity. However, in natural waters, most of the acidity is present due to the dissolution of carbon-dioxide which forms carbonic acid. Determination of acidity is significant as it causes corrosion and influences the chemical and biochemical reactions.

Ammonia

Ammonia of mineral origin is rare in natural waters. The most important source of ammonia is the ammonification of organic matter. Occurrence of ammonia in the waters can be accepted as the chemical evidence of organic pollution.

Phosphates

The major sources of phosphorous are domestic sewage, detergents, agricultural effluents with fertilizers, and industrial waste waters. The prime concern of phosphorous lies in the ability to increase the nuisance algae, and eutrophication.

Sulphate

Sulphate is naturally occurring anion in all kinds of natural waters. It is an important constituent of hardness with calcium and magnesium. Sulphate produces an objectionable taste at 300-400 mg/1. Above 500 mg/1, a bitter taste is produced in the water. At concentrations around 1000 mg/1, it is laxative.

Chlorides

Chlorides occur naturally in all types of waters. The most important source of chlorides in the waters is the discharge of domestic sewage.

Guideline values of drinking water are presented in table-4.

Table 4 : Guideline values for chemical constituents and physical characteristics that effect the aesthetic quality of drinking water.

Constituent or Characteristic	Guideline Value
Aluminum	0.2 mg/l
Chloride	250 mg/l
Colour	15 true colour units
Copper	1.0 mg/l
Hardness	500 mg/l (as CaCOo3)
Hydrogen sulphide	Not detectable by consumer
Iron	0.3 mg/l
Manganese	0.1 mg/l
pH	6.5 - 8.5
Sodium	200mg/l
Solids-total dissolved	1000 mg/l
Sulphate	400 mg/l
Taste and Odour	not offensive for most consumers
Turbidity	5 nephelometric turbidity units, preferably < 1 for disinfection efficiency
Zinc	5.0 mg/l

Source : WHO, 1984

Biological parameters

Biological indicators are neglected compared to the chemical methods (James, 1979). Since water contamination in many instances a biological phenomenon, it would appear logical that it ought be measured biologically. This may be due in part to a lack of understanding of biological data but in many cases it is due to the non-quantitative nature of much of the information (James, 1979). Biological indicators show the degree of ecological imbalance that has been caused and the chemical methods measure the concentration of pollutants responsible. Both types of assessment are therefore necessary.

Biological indicators are useful to assess the water quality required for amenity, fishing, recreation and irrigation. These indicators are also useful in assessing the organic pollution, toxic pollution, solids pollution, eutrophication and heated discharges.

The Bacteriological quality of drinking water

The most common and widespread danger associated with drinking water is contamination, either directly or indirectly by sewage, by other wastes, or by human or animal excrement. If such contamination is recent, if among the contributors there are carriers of communicable enteric diseases, some of the living casual agents may be present. The drinking of water so contaminated or its use in the preparation of certain foods may result in further cases of infection.

Waterborn bacterial pathogens

Faecal pollution of drinking water may introduce a variety of intestinal pathogens-bacterial, viral and parasitic-their presence being related to microbial diseases and carriers, present at that movement in the community. The known intestinal bacterial pathogens are Salmonella, Shigella, enterotoxigenic, Escherichia Coli, Vibrio cholerae, Yersinia enterocolitica and Campylobacter fetus. These organisms may cause diseases that vary in severity from mild gasteroentites to severe and sometimes fatal dysentery, cholera or typhoid. Other organisms such as pseudomonas Flavobacterium, Acinetobacter, Klebsiella and Serratia may produce a variety of infections involving the skin and mucous membranes of the eye, ear, nose and throat (WHO, 1984).

Organisms used as bacterial indicators of faecal pollution include the coliform group of organisms as a whole, E.Coil and Coliform organisms that have been described as 'Faecal Coliforms', faecal streptococci, and sulphite reducing clostridia, especially clostridium ferfringens.

Nuisance Organisms

These constitute a morphologically and physiologically diverse group of organisms, which includes planktonic and sessile algae, fungi, crustacea, and protozoa as well as actinomycetes and iron and sulphur bacteria (WHO, 1984). These organisms can cause objectioable tastes, colour, odour, and turbidity and may interfere with treatment processes by blocking stainers and filters. In addition, certain planktonic organisms, although not themselves harmful, may harbour pathogens and protect them against disinfection by chlorine.

Nuisance organisms may also cause problems with groundwater sources by encrusting well-screens, thus causing loss of yield and impairing the asthetic quality of the supply. Indeed, their presence may indicate organic pollution of the aquifer (Taylor, 1960).

Virological quality of drinking water

Viruses enter the water environment primarily by way of sewage discharges. It is generally believed that the primary root of exposure to enteric viruses is by direct contact with inflected persons or by contact with faecally contaminated objects explosive out breaks of viral hepatites and gastro enterites resulting from sewage contamination of water supplies have been well documented epidemiologically (Craun and McCabe, 1973). Enteric viruses are capable of producing a wide variety of syndromes including rashes, fever, gastroentrities, myocardities, meaningities, respiratory disease and hepatitis.

Free-living organisms

Plankton and macroinvertebrates are generally considered significant in water supplies. Phytoplankton consists of free- living bacteria, fungi and algae. Zooplankton consists of free- living protozoa, rotifers, cladocera, copepods, worms and the larvae of some aquatic insects and fish. The macroinvertebrates include the benthic animals, such as aquatic insect larvae, crustacea and gastropods. These organisms are important in water supplies because they interfere with water treatment processes, produce substances toxic to humans, harbour human pathogens, and contribute organic matter. On the other hand the macroinvertebrates may affect the efficiency of water distribution systems and the acceptance of water by consumers.

Poisonous species occur in yellow-green (Xanthophyta) and blue green (Cyanophyta) algae groups. Algae in water supplies may be aesthetically unpleasant and may interfere with water treatment.

Radiation and Radioactivity

The radionuclides that occur in water are of three sources :

(i) Those that occur naturally - The light nuclides ^{12}C and ^{40}K and the heavy radio nuclides of Uranium (^{238}U, ^{235}U and ^{232}Th).

(ii) The radioactive daughter products of the heavy nuclides.

(iii) Fission and transmutation products that are the result of man's activities.

In many waters it is easy to identify such isotopes as 3_H, 14_C, 40_K and the daughter products of Th and U. The three natural decay series are common (Lee et al., 1977) in the earth's crust and account for a large part of the radiation present and found in water, especially in the form of the three short lived species of radon gas, 219_{Rn}, 220_{Rn} and 223_{Rn}. A short lived isotope of Rn is present in each of the decay chains of 238_U, 235_U and 232_{Th}. The decay products, of these are introduced into water via man's activities (such as cooling water from nuclear reactors). Several species from this source can be routed to man via the pathway of drinking water or aquatic biota. Among these are 3_H' 137_{Cs} and several isotopes of transition metals, such as Fe, Co, Zn and Mn. Lee et al., 1977 have presented on overall water exposure dose equation covering this concern namely

$$D_W = Q/M\,[I_W\, F(DF)\,(e\text{-}\,\lambda ti)\,(ICF + I_s CF_s)\,]$$

Where Q = release (μCiyr^{-1})

M = mass flow of receiving waters (Cm^3 yr^{-1})

t_i = time between emission and exposure (days)

DF = stream dilution factor upon entrance to pathway

I_W = drinking water consumption rate (Cm^3 yr^{-1})

I_s and I = intake of shell fish and fish respectively (Cm^3 yr^{-1})

R = food preparation loss factor

F = water treatment loss factor and

CF_s and CF = shellfish and fish concentration factors respectively.

Somayajulu et al., reported the following concentrations in the drinking water (Table-5)

Table 5 : U isotopes in the drinking water of Ahmedabad After Somayajulu et al.

S No.	Location	^{238}U (μg/l)	$^{234}U/^{238}U$ (Activity Ratio)
1	PRL	20.0 ± 0.3	1.47 ± 0.02
2	PRL	19.2 ± 0.5	1.45 ± 0.01
3	PRL	22.4 ± 0.5	1.42 ± 0.03
4	Vikas Society	21.0 ± 1.5	1.43 ± 0.01
5	Azad Society	13.4 ± 0.8	1.41 ± 0.02
6	Calico Mills Area	12.3 ± 0.6	1.46 ± 0.01
7	Sarangpur	15.4 ± 0.9	1.41 ± 0.02
8	Maninagar	14.2 ± 0.08	1.29 ± 0.02
9	Naroda	12.6 ± 0.8	1.21 ± 0.02
10	Sabarmati River	0.8 – 4.2	1.33 – 1.88

Somayajulu et al.,

Limitations of the water quality criteria

The limitations of water quality criteria (Culp et al., 1980) are listed below:

(a) There is a lack of scientific base from which to establish numerical values for criteria.

(b) There are differing degrees of scientific base from which to establish criteria.

(c) There are differing factors of safety due to the lack of scientific data

(d) Reasons other than public health are used to establish criteria.

(e) The ability to measure low criteria concentrations is not always available.

(f) Continuing scientific research can result in changed criteria.

(g) Other limitations include political and multi-jurisdictional responsibilities.

Water Quality Criteria for Drinking

Health-Related Inorganic Constituents (see table-1b)

Arsenic : Arsenic minerals are widely distributed and small amounts of the free element have also been found. Common minerals include two sulphides, realgar and orpiment and the oxidized form arsenolite.

Many arsenic compounds are water soluble and thus contamination of water can occur. Most of the arsenic found in water derives from industrial discharges. Arsenic is present in most food stuffs.

Acute poisoning by arsenic involves the central nervous system leading to coma and for doses of 70 - 180 mg/l to death (Vallei et al.,1960). The gastro intestinal tract, nervous system, the respirated tract and skin can be severely effected (QMSS,1979).

Asbestos : Asbestos is introduced into natural waters by the dissolution of asbestos containing minerals and ores and from industrial effluents. The use of asbestos - cement pipes in distribution systems is a potential source of asbestos contamination of drinking water. Asbestos is derived from two large groups of rockforming minerals - the serpentines and the amphiboles. The epidemiological studies have shown a marginally significant association between asbestos levels in drinking water and cancers of the digestive tract (McCabe and Millette, 1979).

Occupational exposure to airborne asbestos has resulted in pulmonary fibrosis (asbestosis), pleural calcification, bronchogenic carcinoma of the lung, malignant mesotheliorna of pleura and peritoneum, and cancer of the gastrointestinal tract. Asbestos has been described as a 'chronic inflammatory reaction in the terminal bronchioles and alveoli of the lung, with considerable fibrosis, leading to distortion and eventual obliteration of the alveoli (Davies et al., 1974).

Barium : Most waters contain some barium. The source is normally natural mineral matter (Barite and Witherite). Barium is present in traces in many food stuffs (USEPA, 1976). Brazil nuts are an especially rich source.

Barium is acutely toxic when soluble salts are ingested in excess: if taken as the chloride, the fatal dose of barium for an adult is about 550 - 600 mg/l (NRC, 1977).

Cadmium : Cadmium is found as greenockite. But the commercially important source is found in most zinc ores. Drinking water normally contains very low concentrations of cadmium. Most food stuffs contain traces of cadmium. Higher levels of cadmium in tap water are associated with plate plumbing fittings, silver base solders and galvanized iron piping material (CEC, 1978).

Acute effects have been seen where food has been contaminated by cadmium from plated vessels, severe gastrointestinal upsets have been reported (CEC, 1978: NRC, 1977). Health effects have been demonstrated in industrial workers having exposed to cadmium oxide fumes and dust (CEC, 1979). Bronchitis, emphysema, anemia and renal stones have been reported (NRC, 1977). The renal cortex is generally accepted to be the critical organ for cadmium accumulation in man (WHO, 1972; CEC, 1979; USEPA, 1977). A guideline value of 0.005 mg of cadmium per litre is recommended.

Chromium : Most rocks and soil contain small amounts of chromium. Because of the low solubility of chromium generally, the levels found in water are usually low (NRC, 1974). However, there are examples of contamination of water, in some cases serious, in which effluents containing chromium compounds have been discharged to rivers. The only ore of chromium of any commercial importance is chromite. Other less plentiful sources are crocoite and chrome ochre.

Chromium is absorbed through both the gastrointestinal and respiratory tracts, the amount absorbed differs in each system and depends on the form of chromium (Towill et al., 1978). Chromium appears to be necessary for glucose and lipid metabolism and for utilization of amino acids in several systems. It also appears to be important in the prevention of mild diabetes and atherosclerosis in humans (Towill et al., 1978). The harmful effect of water borne chromium in man are associated with hexavalent chromium. Hexavalent chromium at 10 mg/Kg of body weight will result in liver necrosis, nephritis and death in man; lower doses will cause irritation of the gastrointestinal mucosa (Kaufman et al., 1970). A guideline value of 50 μg/l is proposed.

Cyanide: Cyanide is present wherever life and industry occur. Cyanides form part of life processes, in particular as metabolic intermediates.

Hydrocyanic acid dissociates to gives the cyanide ion in water (USEPA, 1976: QMSS, 1979).

A single dose of 50 - 60 mg for a human being is usually fatal (USEPA, 1976). On the basis of animal experiments it has been calculated that an acceptable daily intake for man is 8.4 mg of cyanide (USEPA, 1979). A guideline value of 0.1 mg/l is recommended.

Fluoride : Fluorine has been fairly conclusively demonstrated to be an essential element for some animal species (Underwood, 1977). The three most important fluorine bearing minerals are fluorite, cryolite, fluorapatite. In little quantities it can reduce dental caries. Once fluoride is incorporated into teeth, it reduces the solubility of the enamel under acidic conditions and there by provides protection against dental caries. Long term consumption of water containing 1 mg of fluoride per litre may lead to mottling (NRC, 1977). Skeletal fluorosis has been observed in persons when water contains more than 3 - 6 mg of F^- /liter. In high doses, F^- is actually toxic to man. Pathological changes include haemorrhagic gastroentrities, acute toxic nephritis and various degrees of injury to the liver and heart muscle (WHO, 1970).

Hardness : Hardness of the water is the property attributable to the presence of alkaline earths (Brown et al., 1970). The major alkaline earth elements present in most natural waters are Ca and Mg. Other elements such as Sr, Ba (also alkaline earths), Fe, Mn and Al are normally not present in waters in sufficient amount to affect a test for hardness.

Many definitions of hardness exist world-wide (Holl, 1972 ; Angino, 1979). The degree of hardness in water is commonly based on the classification (Table-6).

Table 6. Hardness Classification of water (after Sawyer and McCarty)

Hardness, mg/l as $CaCO_3$	Water Class
0–75	Soft
75–150	Moderately hard
150–300	Hard
over 300	Very hard

Water hardness is primarily the result of interaction between water and the geological formations containing it, or over which the water flows. There is some suggestive evidence that drinking extremely hard water might lead to an increased incidence of Urolithiasis (Bokina et al., 1965).

Iron : Iron is abundant in the crust by weight and is a major constituent in pyroxenes, amphiboles, micas (silicates), pyrite and chalco-pyrite (sulphides), and magnetite and haematite (oxides).

Iron is an essential element in human nutrition (Watt and Merrill, 1963). Iron ingestion in large quantities results in a condition haemochromatosis (normal regulatory mechanisms do not operate effectively). Iron bearing water often tastes unpalatable and stains laundry and plumbing fixtures (WHO, 1984).

Lead : The most important lead ore is galena. Other ore minerals are anglesite, cerussite, pyromorphite and mimetestite. Lead in high doses has been recognized for centuries as a cumulative general metabolic poison. Some of the symptoms of acute poisoning are tiredness, lassitude, slight abdominal discomfort, irritability, anaemia and in the case of children, behavioural changes (WHO, 1977).

Mercury : Natural degassing of the earth's crust is major source of mercury. In most surface waters mercuric hydroxide and chloride are the predominant mercury species.

The major effects of mercury poisoning take the form of neurological and renal disturbances (SEG, 1971). In addition to producing general toxic effects, mercury causes gonadotoxic and mutagenic effects and disturbs the cholesterol metabolism.

Nickel : Many nickel salts are water soluble, therefore contamination of water can arise. Garnierite, nickeliferous limonite and pentlandite are the important source minerals.

Nickel is relatively non-toxic element.The levels of nickel usually found in food and water are not considered a serious health hazard (CEC, 1979; Underwood, 1977) , however high doses (1600 mg/kg in the diet) were shown in early animal studies to cause mineral toxic effects (Underwood, 1977).

Nitrate and Nitrite : Nitrates are widely present in substantial quantities in soil, in most waters, and in plants, including vegetables (WHO, 1978).

It has been well documented that, in some countries, water supplies containing high levels of nitrate have been responsible for cases of infantile methaemoglobinaemia and death (QMSS, 1980, USEPA, 1977). This problem does not arise in adults. There are conflicting conclusions regarding the nitrate threshold level for an effect (Frazer and Chilvers, 1980).

Selenium : Selenium has been identified as an essential nutrient in several animal species (WHO, 1973; Schwartz et al, 1961). It is comparatively a rare element. Selenium is essential for human health. Recent studies on Keshan disease suggest that this myocardial disease of children (Xiaoshu chem et al., 1980) could be induced below-level salenium intake. Exposure to high salenium result Nausea, dermatitis and pathological changes in the nails.

Silver : Silver is widely distributed in sulphide ores of which argentite is the most important. Levels of silver in natural waters are very low (QMSS, 1980). There is no evidence that silver is essential to the human organism.

Sodium : Sodium is present mainly in rock salt. This ion is ubiquitous in water owing to the high solubility.

In general, sodium salts are not acutely toxic substances because of the efficiency with which mature kidneys excrete sodium (QMSS, 1980). Excessive intake of sodium chloride causes vomting and the elimination of the much of the salt. Acute effects may include convolutions, muscular twitching and rigidity, and cerebral and pulmonary Oedema (QMSS, 1980). Acute effects and death have been reported in the case of children.

Sulphate : Majority of the sulphates are soluble in water (exceptions, sulphates of lead, barium, and strontium (Mc Kee and Wolf, 1963). Sulphate is poorly absorbed from the human intestine (Novikov and Erisman, 1975). Sulphate doses of 1.0 - 2.0 g have a cathartic effect on humans, resulting in the purgation of the alimentary canal (Mc Kee and Wolf, 1963). Infants ingesting sulphate equivalent to 21 mg/Kg of body weight per day may also suffer from this effect.

High sulphate concentrations in water may contribute to the corrosion of metals in the distribution system, particularly in waters having low alkalinity (WHO, 1984).

Zinc : The major source of Zinc is sphalerite, smithsonite, hemimorphite and franklinite.

Zinc is an essential element for both animals and man and is necessary for the functioning of various enzyme systems (Parisic and

Vallee, 1969).

Symptoms of zinc toxicity in humans include vomiting, dehydration, electrolyte imbalance, abdominal pain, nausea, lethargy, dizziness, and lack of muscular co-ordination (Prasad and Oberleas, 1976). Acute renal failure caused by zinc chloride has been reported (Csata et al., 1968).

Health-Related Organic Contaminants

More than 2000 chemical contaminants of all kinds have been found in water, about 750 of which have been identified in drinking water, of these, more than 600 are organic substances, including many that are pharmacologically active, several that are recognized carcinogen or carcinogenic promoters, and a number that have been shown to be mutagenic.

Compounds of potential health significance, guideline values for health related organic contaminants are shown in the table-7.

Table 7 : Groups of Organic compounds of potential health significance

1. Source contaminants

humic substances

chlorinated alkanes and alkenes

nitrosamines

polynuclear aromatic hydrocarbons (PAH)

nitriolotriacetic acid (NTA)

phenols

synthetic detergents

pesticides

polychlorinated byphenols (PCB)

phthalate esters

petroleum oils, including gasoline

chlorobenzenes

chlorinated phenols

benzene and alkylaromatics

carbon tetrachloride

2. Introduced during treatment

carbon tetrachloride

acrylamide

trihalomethanes

3. Introduced during distribution

vinyl chloride monomer

polynuclear aromatic hydrocarbons (PAH)

Source : WHO, 1984

Quality criteria for livestock and wildlife watering

The permissible daily intake of substances is greatly dependent on the concentrations of the substances and the quality of water ingested. The daily water requirements of animals vary with a number of factors such as temperature and humidity, the water content of the food, the degree of exertion of the animal and the salinity of the water supply (Culp et al., 1980). Therefore, the recommended concentrations of specific substances are based on typical usage.

The more critical parameter of concern is the salinity of the water. Excessively saline waters can cause physiological upset and death of livestock due to osmatic imbalance (Culp et al., 1980). Table-8 shows the range of water salinity and their effect on animals.

Table 8. : Electrical conductivity (EC) guide for drinking water for livestock and poultry.

Total soluble salts Content of waters (mg l^{-1})	Comment
Less than 1000 mg l^{-1} (EC < 1.5 mS cm^{-1})	Relatively low level of salinity. Excellent for all classes of livestock and poultry.
1000-3000mg l^{-1} (EC = 1.5-5 mS cm^{-1})	Very satisfactory for all classes of livestock and poultry. May cause temporary and mild diarrhoea in livestock not accustomed to it or watery droppings in poultry.
3000-5000mg.l^{-1} (EC = 5-8 mS cm^{-1})	Satisfactory for livestock, but may cause temporary diarrhoea or be refused at first by animals not accustomed to it. Poor water for poutlry, often causing water faeces, increased mortaltiy, and decreased growth, especially in turkeys.
5000-7000 mg l^{-} (EC = 8-11 mS cm^{-1})	Can be used with reasonable safety for dairy and beef cattle, sheep, swine, and horses. Avoid use for pregnant or lactating animals. Not acceptable for poultry.
7000-10,000 mg l^{-1} (EC = 11 - 16 mS cm^{-1})	Unfit for poultry and probably for swine. Considerable risk in using for pregnant or lactating cows, horses, or sheep, or for the young of these species. In general, use should be avoided although older ruminants, horses, and swine may subsist on them under certain conditions.
Over 10,000 mg l^{-1} (EC > 16 mS cm^{-1})	Risks with these highly saline waters are so great that they cannot be recommended for use under any conditions.

After National Academy of Sciences and National Academy of Engineering 1972.

The concentrations of toxic substances which render a water

undesirable are dependent on many variables. These include age, sex, species, physiological state of animal and the factors mentioned before. The recommended standard levels are shown in table-9.

Table 9 : Recommendations for levels of toxic substances in drinking water for livestock

Constituent	Upper limit (mg l^{-1})
Aluminium (Al)	5
Arsenic (As)	0.2
Beryllium (Be)	No data
Boron (B)	5.0
Cadmium (Cd)	0.05
Chromium (Cr)	1.0
Cobalt (Co)	1.0
Copper (Cu)	0.5
Fluoride (F)	2.0
Iron (Fe)	No data
Lead (Pb)	0.1[a]
Manganese (Mn)	No data
Mercury (Hg)	0.01
Molybdenum (Mo)	No data
Nitrate + Nitrite (NO_2-N+NO_2-N)	100
Nitrite (NO_3-N)	10
Selenium (Se)	0.5
Vanadium (V)	0.10
Zinc (Zn)	24
TDS	10 000

After National Academy of Sciences and National Academy of Engineering 1972.

[a] Lead is cumulative and problems may begin at a threshold value of 0.05 mg l^{-1}

Water Quality for Irrigation

The required quality of irrigational waters varies substantially, depending principally upon the salinity, soil permeability, toxicity and some miscellaneous concerns such as excessive nitrogen loading or unusual pH of water. The recommended water quality criteria for irrigation is shown in table-10.

Table 10 : Summary of Water Quality Criteria for Agricultural Irrigation

Parameter	National Academy of Science, 1972	McKee and Wolf 1963	U.S. EPA, Oct. 1973	
			Long Term	Short Term
Aluminum	5.0	1.0	5.0	20.0
Arsenic	0.1	-	0.1	2.0
Bacteria				
Fecal Coliform	1000/100 ml	-	1000/100 ml	-
Beryllium	0.1	-	0.1	0.5
Boron	1	-	-	-
Cadmium	0.01	-	0.01	0.05
Chloride	-	100	0.4-1.8	-
Chromium	0.10	3	0.1	1.0
Cobalt	0.050	0.1	0.05	5.0
Copper	0.2	0.17	0.2	5.0
Fluoride	1.0	10.0	2.0	15.0
Iron	5.0	-	5.0	20.0
Lead	5.0	-	5.0	10.0
Lithium	2.5	0.05	2.5	-
Manganese	0.2	0.5	0.2	10.0
Molybdenum	0.01	-	0.01	0.05
Nickel	0.2	0.5	0.2	2.0
pH	4.5-9.0	6.0-9.0	-	-
Phenols	-	50	-	-
Selenium	0.02	0.0-0.1	0.02	-
SAR	1	-	1	-
Solids				
Total Dissolved	3	-	3	-
Sulphate	-	200	-	-
Vanadium	0.10	0.1	-	-
Zinc	2.0	3.0	-	-

Classification of irrigation waters

Irrigation water is classified on the basis of the following characteristics.

(a) salinity

(b) sodicity

(c) concentration of carbonate and bicarbonate ions

(d) concentration of elements that are toxic to plant growth

(a) Salinity : The salinity or total concentrations of soluble salts is usually measured as electrical conductivity (EC) in irrigation work and the US salinity laboratory has established guide groupings of water based on this parameter (Table-11).

Table 11. Groups of irrigation waters based on electrical conductivity

TDS (mg l^{-1})	EC x 10^6 at 25°C (μS cm^{-1})	Class	Remarks
< 200	< 250	C_1	Low salinity water–can be used for irrigation with most crops on most soils with little likelihood that a salinity problem will develop. Some leaching is required, but this occurs under normal irrigation practices except in soils of extremely low permeability.
200-500	250-750	C_2	Medium salinity water–can be used if a moderate amount of leaching occurs. Plants with moderate salt tolerance can be grown in most instances without special practices for salinity control.
500-1500	750-2250	C_3	High salinity water–cannot be used on soils with restricted drainage, special management for salinity control may be required and plants with good salt tolerance should be selected.
1500-3000	2250-5000	C_4	Very high salinity water–is not suitable for irrigation under ordinary conditions but may be used occasionally under very special circumstances. The soil must be permeable, drainage must be adequate, irrigation water must be applied in excess to provide considerable leaching, and very salt-tolerant crops should be selected.

After U.S. Salinity Laboratory 1954.

(b) Sodicity : Sodicity or sodium hazard may be assessed using Soluble Sodium Percentage (SSP), Sodium Adsorption Ratio (SAR) or Exchangeable Soidum Ratio (ESR). The relavant equations are;

$$SSP = \frac{100\,(Na^+ + K^+)}{(Ca^{++} + Mg^{++} + Na^+ + K^+)}$$

all the concentrations are expressed in meq/l

$$SAR = \frac{Na^+}{\sqrt{(Ca^{++} + Mg^{++})/_2}}$$

all the concentrations are expressed in meq/1

$$ESR = \frac{Na^+}{Ca^{++} + Mg^{++}} = \frac{Na^+}{CEC - Na^+}$$

CEC is the Cation Exchange Capacity and all the concentrations are in meq/100 g of soil.

The relationship between SAR and ESR is given by

ESR = a + b (SAR)

where a and b are empirical constants for a particular soil type and needed to be locally established

Table 12 Irrigation water classification based on sodium content

Classification	Comment
S_1	Low sodium water can be used for irrigation on almost all soils with little danger of the development of harmful levels of exchangeable sodium. However, sodium-sensitive crops such as stonefruit trees and avocado may accumulate injurious concentrations of sodium.
S_2	Medium sodium water will present an appreciable sodium hazard in fine textured soils having high cation exchange capacity, especially under low leaching conditions, unless gypsum is present in the soil. This water can be used on coarse-textured or organic soils with good permeability.
S_3	High sodium water may produce harmful levels of exchangeable sodium in most soils and will require special soil management- good drainage,high leaching, and organic matter additions. Gypsiferous soils may not develop harmful levels of exchangeable sodium from such waters. Chemical amendments may be required for replacement of exchangeable sodium except that amendments may not be feasible with waters of very high salinity.
S_4	Very high sodium water is generally unsatisfactory for irrigation purposes except at low and perhaps medium salinity, where the dissolving of calcium from the soil or the use of gypsum or other additives may make the use of these waters feasible.

After Wilcox 1955.

The sodium hazard of irrigation waters as expressed by SAR does not take into account the effect of anion composition. The empirical equation was Varified by paliwal and Maliwal (1971) with varying amounts of CO_3 and HCO_3 ions and could a low degree of correlation in case of water of Residual Sodium Carbonate more than 2.0 meq/1. Soil properties deteorated with increase in SAR levels in irrigation water

(Sreerama Murthy, 1989). A general classification of water based on sodium concentration is given in table-12 and the irrigation water classification adopted by the US Department of Agriculture based on SAR in combination with EC is shown in fig. 1. Kanwar (1961) modified the classification with wider range of EC and added one more class from 5 to 20 mmhos/cm. Ramamoorthy (1964) also modified the classification in the higher range C_4 from 2.25 to 6.75 and C_5 from 6.75 to 20.25 mmhos/cm.

(c) Concentration of carbonate and bicarbonate ions : To quantify the effects of carbonate and bicarbonate, Eaton (1950) proposed Residual Sodium Carbonate (RSC), which is given by

$$RSC = (CO_3^- + HCO_3) - (Ca^{++} + Mg^{++})$$

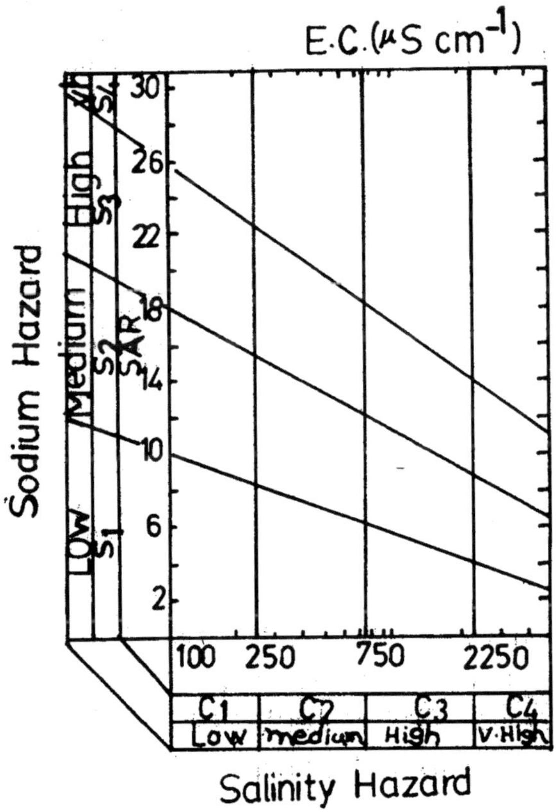

Fig. 1. Salinity and sodium adsorption classification for irrigation water use.

and is in units of meq/1. The guidelines for applying RSC are given in table-13.

Table-13 Guidelines for applying RSC values

Condition	RSC (meq/1)
Suitable	< 1.25
Marginal	1.25 – 2.5
Not suitable	> 2.5

(d) Toxic elements

1. Boron : Boron is essential for plant growth however, it is extremely toxic at concentrations slightly above optimum. The criteria adopted are given in tables-14 and 15.

Table 14 Permissible limits of boron in irrigation waters (mg l^{-1})

Class of water	Limit (mg l^{-1})		
	Sensitive crops	Semi-tolerant crops	Tolerant crops
Excellent	<0.33	<0.67	<1.00
Good	0.33–0.67	0.67-1.33	1.00–2.00
Permissible	0.67–1.00	1.33–2.00	2.00–3.00
Doubtful	1.00–1.25	2.00–2.50	3.00–3.75
Unsuitable	>1.25	>2.50	>3.75

Source : Wilcox, 1960

Table 15 Relative tolerance of plants to boron

Sensitive	Semi-tolerant	Tolerant
Lemon	Lima bean	Carrot
Grapefruit	Sweet potato	Lettuce
Advocado	Bell pepper	Cabbage
Orange	Pumpkin	Turnip
Apricot	Zinnia	Onion
Peach	Oat	Broadbean
Cherry	Milo	Gladiolus
Persimmon	Corn	Alfalfa
Kadota fig	Wheat	Gardenbeet
Grape	Barley	Mangel
Apple	Olive	Sugarbeet
Pear	Field pea	Date palm
Plum	Radish	Asparagus
Navy bean	Sweetpea	Athel (Tamarix aphylla)
Jerusalem artichoke	Tomato	
Walnut	Cotton	
Pecan	Potato	
	Sunflower	

Source : Richardson (1954)

2. Magnesium Hazard : Szabolcs and Darab (1964) proposed Magnesium Hazard (MH) for irrigation water and is given by

$$MH = \frac{Mg^{++}}{Ca^{++} + Mg^{++}} \times 100$$

The units are in meq/1, and where MH > 50 the effects are considered to be harmful.

3. Trace elements : Branson et al. (1975) have studied the plant tolerance to trace elements. The table summarized by Lloyd and Heathcote is included here. (Table - 16).

Table 16 Recommended maximum concentration of trace elements in irrigation water[a]

Elements	For waters used continuously on all soil ($mg\ l^{-1}$)	For use up to 20 years of fine-textured soils at pH 6.0 to 8.5 ($mg\ l^{-1}$)
Aluminium	5.0	20.0
Arsenic	0.10	2.0
Beryllium	0.10	0.50
Boron	0.75	2.0–10.0
Cadmium	0.010	0.050
Chromium	0.10	1.0
Cobalt	0.050	5.0
Copper	0.20	5.00
Fluorine	1.0	15.0
Iron	5.0	20.0
Lead	5.0	10.0
Lithium	2.5	2.5 [b]
Manganese	0.20	10.0
Molybdenum	0.010	0.050 [c]
Nickel	0.20	2.0
Selenium	0.020	0.020
Vanadium	0.10	1.0
Zinc	2.0	10.0

After Branson et al., 1975.

[a] These levels will not normally have an adverse effect on plants or soils. No data are available for mercury, silver, tin, titanium, or tungsten.

[b] Recommended maximum concentration for citrus is 0.75 $mg\ l^{-1}$

[c] Only for fine-textured acid soils, or acid soils with a relatively high content of iron oxide.

Crop Tolerance and Water Quality

Crop tolerance to salt stress has been assessed on the basis of soil salinity (Maas and Hoffman, 1977; Maas, 1986) soil exchangeable sodium percentage (Pearson and Bernstein, 1958; Gupta and Sharma, 1990), specific ions such as boron (Maas, 1986) and sodium (Maas, 1986; Roads, 1972). Maas and Hoffman (1977) visualized and proposed a model for describing crop tolerance which states that crops do not suffer yield reduction unless a threshold value is reached beyond which yield declines linearly with increasing soil salinity. Prichard et al. (1983) suggested that for estimating suitability of a crop to irrigation with saline water; information on its tolerance to soil salinity should be known and a relation between soil salinity and irrigation water quality need to be developed. Such relations depended upon a number of factors and therefore, are locations specific and at the same location vary from an year to another (Prichard et al., 1983). Similar limitations restrict the direct use of groundwater quality in the model. Gupta (1994) has treated the irrigation water quality as a variable directly for assessing crop response to irrigations with saline waters.

Piece wise linear model of Maas and Hoffman (1977) is written as

$$Y = \begin{matrix} 100 & O < EC_{iw} \leq EC_t \\ 100 - S(EC_{iw} - EC_{t)} & EC_t < EC_{iw} \leq ECO \\ 0 & EC_{iw} > ECO \end{matrix}$$

Here Y is the relative yield with respect to conditions where non-saline water has been used for irrigations:EC_{iw} is the quality of irrigation water in terms of its electrical conductivity; EC_t is the electrical conductivity at or below which there is no yield reduction compared to non - saline water and S is the slope. The slope represents the yield reduction per unit increase in the electrical conductivity of the irrigation water. The parameter ECO is the electrical conductivity at or beyond which yield will be practically zero. Irrigation water quality at which pre-determined decrease in yield will be practically zero. Irrigation water quality at which pre- determined decrease in yield will occur can be assessed by the relation (Gupta, 1994).

$$EC_{iw} = EC_t + \frac{\text{Desired yield reduction (\%)}}{S\ (\%)}$$

Water quality for Industry

It is difficult to summarize the water quality requirements for industry. The requirements vary considerably between countries, industries and process, and will be specific to a particular product. Suggested water quality tolerances for various industries are shown in tables 17 - 22.

Table 17. Summary of water quality criteria for chemical and allied products (All units in mg/l except as noted)

Parameter	National Academy of Science, 1972	National Technical Advisory comm., 1968	Mckee and Wolf, 1963		
			Production	Manufacture	AWWA 1950 Plastics
Alkalinity	500	500	50-75	-	-
Aluminum	-	-	8	-	-
Bicarbonte	600	600	-	-	-
BOD	1	1	-	-	-
Calcium	250	200	-	-	-
Chloride	500	500	-	-	-
COD	1	1	-	-	-
Color, Units	500	500	5	-	2
Copper	-	-	5	-	-
Hardness	1,000	1,000	8	55	-
Iron	10	5	0.05	0.0	0.02
Iron and Manganese	-	-	0.05	0.0	0.02
Magnesium	100	100	-	-	-
Manganese	2	2	0.03	0.0	0.02
Nitrogen Ammonia	1	1	-	-	-
Odor	1	1	-	-	-
Oxygen, Dissolved	1	1	-	-	-
pH, Units	5.9-9.0	5.5-9.0	-	7.8-8.3	-
Silica	1	1	-	-	-
Solids Suspended	10,000	10,000	-	-	-
Total Dissolved	2,500	2,500	-	-	200
Sulfate	850	850	-	-	-
Temperature	1	1	-	-	-
Turbidity	-	-	5.0	0.3	2

Table 18. Summary of water quality criteria for petroleum and allied products (All units in mg/l except as noted)

Parameter	National Academy of Science, 1972	National Technical Advisory Comm., 1968	Amer. Petrol- Inst.
Alkalinity	500	-	300
Bicarbonate	480	480	-
Calcium	220	220	60-80
Chloride	1,600	1,600	100-300
COD	1,000	-	-
Color, units	25	25	-
Fluoride	1.2	1.2	-
Hardness	900	900	-
Iron	15	15	-
Iron and Manganese	-	-	0.3
Magnesium	85	85	60-80
Nitrogen			
Ammonia (as N)	10	-	-
Nitrate (as N)	8	8	-
Oxygen, dissolved	-	-	0.007
pH, units	6.0-9.0	6.0-9.0	-
Silica	85	50	3-20
Sodium	-	230	Low as possible
Sodium & Potassium	230	-	-
Solids			
Suspended	5,000	5,000	5
Total Dissolved	3,500	3,500	-
Sulfate	900	570	100-300
Sulfide (H_2S)	20	-	0.5

Table 19. Summary of Water Quality Criteria For Primary Metals Industries (All units in mg/l except as noted)

Parameter	National Academy of Science, 1972	McKee and Wolf 1963	Bethlehem Steel Corp.
Alkalinity	2	-	-
Bacteria Total Coliform	-	-	$5x10^6$ MPN/100 ml
BOD	-	-	46
Chlorides	3	175	100
Hardness	100	50	-
Heavy metals	-	-	Trace
Oil	1.0	-	-
Organics	-	Low as possible	-
Oxygen, Dissolved	Aerobic	-	-
Nitrogen Nitrate	-	-	4
pH	6.0-9.0	6.8-7.0	7.0
Phosphate	-	-	12
Sodium	-	-	75
Solids			
Settleable	5	-	-
Suspended	5-10	25	44
Total Dissolved	3	-	450
Temperature °F	100	75	-

Table 21. Summary of water quality criteria for paper and allied products (All units in mg/l except as noted)

Parameter	Water Quality Recommended by TAPPI				
	Fine Paper	Kraft Paper		Ground wood Papers	Soda and Sulfite Pulp
		Bleached	Unbleached		
Alkalinity	75	75	150	150	75
Calcium hardness	50	-	-	-	50
Carbon dioxide, free	10	10	10	10	10
Chlorides	-	200	200	75	75
Chlorine, residual	2.0	-	-	-	-
Hardness, total	100	100	200	200	100
Iron	0.1	0.2	1.0	0.3	0.1
Magnesium hardness	-	-	-	-	5.0
Manganese	0.03	0.1	0.5	0.1	0.05
silica	20	50	100	50	20
Solids Total Dissolved	200	300	500	500	250
Turbidity	10	40	100	50	25

Table 20 : Summary of water quality criteria for food and kindred products (All units in mg/l except as noted)

Parameter	National Academy of Sciences, 1972			McKee and Wolf 1963			Water quality used by British and Australian Industries	
	Food Canning	Fruits & Vegetables	Soft Drink	General	Equip washing	Food canning	WHO[2]	USPHS[4]
Acidity	-	0	-	-	-	-	-	-
Alkalinity	300	250	85	30-250	-	30-250	-	-
Arsenic	-	-	-	-	-	-	0.5	0.05
Bacteria, MPN / 100 ml								
Total Coliform	5	5	-	-	-	-	-	-
Barium	-	-	-	-	-	-	1.0	1.0
Cadmium	-	-	-	-	-	-	.01	0.01
Calcium	12[illegible]	100	-	-	-	-	75	-
COD	6	-	-	-	-	-	-	-
Chloride	300	250	500	-	250	1000-1500	200	250
Chlorine	-	7	-	-	-	-	-	-
Chromium	-	-	-	-	-	-	.05	.05
Color, P-C Units	5	5	5	5-10	5-20	-	5	15
Copper	-	-	-	-	-	-	.1	1.0
Cyanide	-	-	-	-	-	-	-	0.01
Hardness	310	250	8	10-250	10	25-40	1.0	-
Fluoride	5	1	-	1.0	1.0	1.0	1.0	9
Iron	0.4	0.2	0.3	0.2	-	0.2	0.3	0.3
Iron and Manganese	-	-	-	0.2-0.3	0.2	0.2-0.3	-	-

1	2	3	4	5	6	7	8	9
Lead	–	–	–	–	–	–	.05	0.05
Magnasium	–	–	–	–	–	–	50	–
Manganese	0.2	0.2	0.05	0.2	–	0.2	0.1	0.05
Nitrogen								
Ammonia (as N)	–	–	–	–	–	0.6	–	–
Nitrate (as N)	10	10	–	–	–	2.0	10	10
Nitrite (as N)	0	0	–	–	–	–	–	–
Odor Units	5	0	–	0	0	0	unobjec-tionable	3
Organics,								
CTE	–	0.2	–	–	Trace	–	–	–
CCE	0.3	–	0.2	–	Trace	–	0.2	0.2
pH, Units	8.5	6.5.8.5	8	–	–	77.5	7.0-8.5	–
Phenols	5	–	–	–	–	–	0.001	0.001
Salenium	–	–	–	–	–	–	0.01	0.01
Silica	50	50	–	–	–	–	–	–
Silver	–	–	–	–	–	–	–	0.05
Solids,								
Suspended	12	10	–	–	–	–	–	–
Total Dissolved	550	500	8	850	850	850	–	–
Sulfate	250	250	500	–	–	–	250	250
Sulfide	–	–	–	–	–	1.0	–	–
Taste	10	0	–	0	0	0	unobjec-tionable	–
Turbidity	–	5	–	1-10	1.0	1.0-10.0	5	3
Zinc	–	–	–	–	–	–	5	5

Table 22 : Summary of water quality criteria for boiler feed water (All units in mg/l except as noted)

Parameter	Mckee and Wolf 1963				EPA, 1975 Range of Values (above 2000 psig)	Natural Academy of Sciences, 1972		Water Quality Criteria Used for This Evaluation			
	0-150	150-250	250-400	Over 400		Indus. (psig) 0-1500	Util (psig) 700-5000	Indus (psig) 0-150	Indus (psig) 150-700	Indus (psig) 700-1500	Util. (psig) 1500-5000
Alkalinity	-	-	-	-	-	500	500	150	100	40	1
Aluminum (oxide)	5	0.5	0.05	0.01	-	3.0	3.0	5	0.1	0.01	0.01
Bicarbonate	50	30	5	0	-	600	600	170	120	48	0.5
Calcium						-	-	3	0.4	0.01	0.01
Carbonate	200	100	40	20	-	-	-				
Chloride	-	-	-	-	0.005-0.100	19,000	19,200	3	3	3	3.4
Color, P-C Units	80	40	5	2		1,200					
Copper	-	-	-	-	0.002-0.010			0.5	0.05	0.05	0.01
COD	-	-	-	-		100	500	5.0	5.0	1.0	1.0
Hardness	80	40	10	2	0	5,000	9,000	150	0.1	0.07	0.07
Hydrogen Sulfide	5	3	0	0	-	-	-	3	3	3	3
Hydroxide	50	40	10	15	-	-	-				
Iron	-	-	-	-	0.005-0.010	80	90	1	0.1	0.05	0.01
Magnesium	-	-	-	-	-	-	-	1	0.25	0.01	0.01
Manganese	-	-	-	-	-	-	10	0.3	0.1	0.01	0.01

Table Contd...

Parameter	Mckee and Wolf 1963				EPA, 1975 Range of Values (above 2000 psig)	Natural Academy of Sciences, 1972		Water Quality Criteria Used for This Evaluation			
	0-150	150-250	250-400	Over 400		Indus. (psig) 0-1500	Util (psig) 700-5000	Indus (psig) 0-150	Indus (psig) 150-700	Indus (psig) 700-1500	Util. (psig) 1500-5000
Nitrogen											
Ammonia (as N)	-	-	-	-	-	-	-	0.1	0.1	0.1	0.07
Organics											
PHAS	-	-	-	-	-	2	10	1	1	0.5	0.1
CTE	-	-	-	-	-	100	100	1	1	0.5	0
Oxygen, Dissolved	1.4	0.14	0.0	0.0	0.005	-	-	2.5	0.007	0.007	0.007
pH (minimum), Units	8.0	8.4	9.0	9.6	6.0-9.6	-	-	7.0-10.0	8.2-10.0	8.2-9.0	8.8-9.4
Phosphate						-	50				
Silica	40	20	5	1	0.005-0.50	150	150	30	10	0.7	0.1
Solids,											
Suspended					0.0-0.050	15,000	15,000	10	5	0.5	0.05
Total Dissolved	3000-500	2500-500	1500-100	0.05-0.50	0.0-0.050	35,000	35,000	800	500	200	0.5
Sulphate						1,400	1,400	3	3	3	4
Temperature, °C					-	120	120	3	3	3	3
Turbidity, TU	20	10	5	1	-	-	-				
Zinc					-	-	-	3	0.01	0.01	0.01

Water Quality and Hydrochemistry

Water can be considered as an aggressive solvent that interacts with exposed geologic media on its way through the landscape. The water will act as a solvent for ionic species, as an active agent in weathering and enrichment processes, as a medium where fast processes and changes of conditions and chemical states can take place, as well as a medium for mass and heat transfer. The resulting water quality reflects the combined contributions from various processes along the water path way.

Some dominating processes and reaction systems that will determine concentrations of major inorganic species in natural fresh waters are illustrated (See Beall and Allard, 1979)

WATER QUALITY AND REMOTE SENSING TECHNIQUE

Groundwater quality can be deciphered using remote sensing techniques. It is possible through delineation of hydro- geomorphic zones in the field. Bhatnagar and Naik (1983) had demarcated four hydro morphic Zones in the coastal plain of Gujarat. They are

(1) Tidal mud flat

(2) Filled lagoon

(3) Old meander belt and

(4) Terrace plain of the river Mahi.

The photocharacteristics of these groundtruth are presented in table-23

Table 23. Quality of groundwater in different hydro-geomorphic zones of Gujarat main land (Bhatnagar and Naik 1983)

Sl. No.	Hydrό geomorphic unit	Inferred quality	Range of specific conductance (μ mhos)
1.	Tidal flat zone	Very bad	10,000
2.	Filled lagoon	Bad to very bad	3000 to 9500
3.	Terrace plain	Marginal	600 to 1700
4.	Old meander belt	Good to very good	2000 to 3500

Natarajan et al., (1983) had also presented the water quality of parts of Metwa area, Haryana with the help of aerial photographs. According to their study, water quality is normal to good in pediment plain, marginally saline to saline sodic in inter mountain basin, normal to good in the Yamuna alluvial plain.

Groundwater Quality Monitoring - Statistical Modeling Multiple linear regression models

The entire water quality can be predicted through a single simple test like specific conductance within turn clearly indicates the total dissolved solids. (Basivi Reddy and Pandu Ranga Rao, 1989).

TDS is the single parameter which could reflect the influence of all the dissolved constituents. It can be shown as

$$TDS = fe' (Na, Ca, K, Mg, Cl, (HCO_3 + CO_3), SO_4, NO_3, F)$$

Multiple linear regression is the most useful model available in the literature. The models containing nitrates, sulphates, and pH were found not to be satisfactory (Basavi Reddy and Pandu Ranga Rao 1989). The recommended models based on multiple regression analysis are presented below.

In order to predict water quality throughout a year, attempts are made for conducting regression analysis on the data from 290 samples collected throughout the year over five seasons and the following models have been found satisfactory.

1. $TDS = 3.25\ Na + 480$
2. $TDS = 2.25\ Cl + 480$
3. $TDS = 3.2\ Na + 18.2K + 340$
4. $TDS = 3.00\ Na + 5.75\ Ca + 140$
5. $TDS = 6.40\ Mg + 2.20\ Cl + 380$
6. $TDS = 2.20\ Cl + 0.35\ HCO_3 + 380$
7. $TDS = 3.15\ Na + 18.5\ Mg + 175$
8. $TDS = 6Mg + 21.5\ Cl + 0.334\ HCO_3 + 295$

On the whole it appears, that sodium either independently or in association with K, Mg, Ca and HCO_3 are the casual variables for the determination of TDS variable. Alternatively chlorides, either independently or in association with bicarbonates and Mg also appear to be the important variables for the contribution of TDS in the water. It is now possible to predict TDS in any water given one predictor value.

One important application of this analysis is that the entire water quality can be predicted through a single simple test like specific conductance which in turn clearly indicates the TDS.

REFERENCES

Angino, E.E. (1979) Geochemistry of water in relation to cardiovascular disease, PP.-13.

Basivi Reddy KSV and Pandu Ranga Rao M., (1989) Groundwater quality monitoring-statistical modeling (1989) Intern, Symp, NGRI, Hyderabad Vol,

II PP.755-759.

Beall, G.W. and Allard, B. (1979) Transactions of nuclear society Annual meeting, 32, PP,243.

Bhatnagar, G.C. and Naik, S.D. (1983) Quality of groundwater as evaluated from remotely sensed data in a part of coastal plain of Gujarat main land, PP.231-237. Proc.Nat.Symp. on Remote sensing techniques in development and management of water resources.

Bokina et al, (1965) In Guidelines for drinking water quality, WHO 1984, 333 PP.

Branson RL Pratt PF Roades JD and Oster JD, (1975) Water quality in irrigated watersheds. Jour. Envi. Quality, 4 PP.33-40.

Brown,E. Snougsted, M.W and Fishmen, M.J. (1970) In techniques of water resources investigations of US Geological Survey, Book 5. US Government printing office, Washington, DC.

Bruvold WH et al., (1967) Consumer attitudes toward mineral taste in domestic water. Jour. Am. Water Works. Assoc. 59, P. 547.

Commission of the European Communities (1978) Criteria (dose/effects relationship) for Cadmium. Oxford, Pergamon press.

Commission of the European Communities (1979) Trace metals exposure and health effects. Oxford, Pergamon press.

Coates MS Haimson CB Hinze WJ and Van Schmuss WR., (1983) Introduction to the Illinois Deep Hole Project. Jour. Geophy. Res. 88,B9.

Coutre RA Steitz MG and Steindler MJ., (1983) Sampling of Brine in Cores of Precambrian granite from Northern Illinois. Jour. Geophys. Res. 88, B9. PP. 7331-7334.

Craun GF and Mc Cabe LJ., (1973) Review of the causes of waterborne disease outbreaks. Jour. of the Am. Water Works Association. 65,pp. 74-84.

Csata S. et al., (1968) in Guidelines for drinking water quality, WHO, 1984,333 PP.

Culp G George Wesner Rober Williams and Mark V Hughes (Jr.) (1980) Waste water reuse and recycling technology, Pollution technology Review, No. 72,838 PP.

Daniels JJ Olhoeft GR and Scott JH., (1983) Interpretation of core and well log physical property data from drill hole VPH-3, Stephenson Country, Illinois. Jour. Geophy. Res. 88 B9, PP. 736- 54.

Davies P et al., (1974) Asbestos induced selective release of lysosomal enzymes from mono-nuclear phagocytes. Nature, 251, P. 423.

Davis SN and Turk LJ., (1964) Optimum depth of wells in crystalline rocks. Ground water, 2(2) PP. 6-11.

Dunigion, A.S. Jr and Crawford, N.H. (1976) Modeling non point pollution from the land surface. U.S. Environmental Protection Agency Rep.600/2-76-043, PP. 33-40.

Eaton FM, (1950) Significance of carbonates in irrigation waters. Soil Sci., 69, PP. 123-33.

Falkenmark M and Allard B., (1991) water quality genesis and distrubances of natural water, Hand book of Environmental chemistry, Ed.O.Hutzinges. V,5, Part-A,Springer-Verlag, 264 PP.

Frazer P and Chilvers C., (1980) Health aspects of nitrate in drinking water,

Netherlands. Inter. symp. on water supply and health.

Farmer, W.J. and Letey, J. (1974) Volatilization losses of pesticides from soils. US. Environmental Protection Agency, Rep. 660/2-74 - 054, PP. 24 -51.

Gupta SK., (1994) Management of poor quality of groundwaters for crop production in India. Inland water Resources: India, Eds. M.K.Durga Prasad and P.Sankara Pitchaiah, Ajanta Publications, New Delhi.

Gupta SK and Sharma SK., (1990) response of crops high Exchangeable Sodium percentage, Irrigation Science.

Haimson BC and Doe TW (1983) State of stress permeability and fractures in the Precambrian granite of Northern Illionis. Jour. Geography. Res. 88 B9, PP: 7355-7371.

Harvey, J.C., (1982) The properties of groundwater. John Willey & Sons, 406 PP.

Hem JD., (1970) Study and interpretation of the chemical characteristics of natural water. U.S. Geol.surv. water supply. Paper. 1473.

Holl, K. (1972) water examination, assessment, conditioning, chemistry, bacteriology,biology Walter de Gruyter, Berlin.

James, A. (1979) The value of biological indicators in relation to other parameters of water quality. In biological indicators of water quality, Eds. A. James and Lilian Evison, John Wiley & Sons New York. SI-28PP.

Kanwar JS., (1961) Quality of irrigation water or an index of its suitability for irrigation purposes, Potash Rev. 18, PP. 1-13.

Karanth KR (1989) Hydrogeology, Tata-Mc Graw Hill, New Delhi, 458 PP.

Kaufman DB et al., (1970) Acute potassium dichromate poisoning in man. AM. Jour. Deseases of children, 119, 374.

Lee, H. Peyton, T.O. Steele, R.V. and White R.K. (1977) Potential radioactive pollutants resulting from expanded energy programs. EPA 600/7-77-082, Environmental protection Agency, Washington, DC.

Liegeois (1928) In the properties of groundwater, J.C. Harvey, 1982, John Wiley & Sons. 406 PP.

Linsley RK Kohler Jr. M.A. and Paulhus JLH., (1988) Hydrology for Engineers. Mc Graw Hill. PP. 412-441.

MacCracken, M.C. and Sauter,G.D. (eds) (1975) Development of an air pollution model for the San Francisco Bay area, Lawrence Livermore Lab, Rep. UCRL-51920.

Maas EV., (1986) Salt tolerance of plants. Applied Agri. Res. 1, PP. 12-26.

Maas EV and Hoffman GI., (1977) Crop-salt tolerance - current assessment. Am. Soc. Civ. Eng. Irrigation and Drainage Div. IR_2, PP. 115-134.

Mc Cabe LJ and Millette Jr., (1979) Health effects and prevalence of asbestos fiber in drinking water. Am. water works Assoc. Ann. Conference, San Francisco, Denver, CO, AWWA, PP. 24-29.

Mc Kee JE and Wolf HW., (1963) Water Quality Criteria 2nd ed., Sacramento, CA, California State Water Quality Control Board.

Natarajan,S. Gajbe,M.Y. and Manchanda, M.L. (1983) Use of aerial photographs for study of groundwater quality of parts of Mewat area, Haryana, pp,217-234, Proc. Nat. Symp. on Remote sensing techniques in development and management of water resources.

National Academy of Sciences and National Academy of Engineering (1972)

Water quality criteria. REP . EPA- R3-73-033. Washington.

National Research Council (1974) Chromium. Washington DC. National Academy of Sciences.

National Research Council (1977) Drinking water and health, Washington DC. National Academy of Sciences.

Novikov Yu and Erisman FF., (1975) The Potable water standard (Gost 2874-73).

Paliwal and Maliwal (1971) In water quality for irrigation. Water ecology, pollution and management, 2, 1991, Pp. 212-231.

Parisic AF and Vallee B.L., (1969) Zinc metalloenzymes, characteristics and significance in biology and medicine. Am. Jour. Clin. nutrition, 22,P.1222.

Pearson GA and Bernstein L., (1958) Influence of exchangeable sodium on yield and chemical composition of plants (2) wheat, barley, oats, rice, tall fescue and tall wheat grass. Soil Sci. 86, PP. 254-259.

Pricard TC Mayer JL Hoffman GJ Krgel FR and Robertrs, R., (1983) Relationship of irrigation water salinity and soil water salinity. California Agriculture:37, PP. 11-14.

Prasad AS and Oberleas D., (1976) Trace elements in human health and disease. Vol. 1. zinc and copper. Academic Press, New York. 470 PP.

Quebec, Ministry of Supply and Services (1979) Guidelines for Canadian drinking water quality.

Quebec, Ministry of Supply and Services (1980) Guidelines for Canadian drinking water quality.

Ramamoorthy B., (1964) Management of soil-water-plant-climate complex with special reference to saline and sodic soils in the Arid zone. Proc. Symp. Probl. Indian Arid Zone, Jodhpure, PP. 143-147.

Roades J.D., (1972) Quality of water for irrigation. Soil Sci., 113, PP.277-284.

Sawyer CN and Mc Carty PL., (1967) Chemistry for Sanitary engineers. 2nd. ed. Mc Graw Hill Co. New York, 518 PP.

Schwarz K et al., (1961) Introduction Symposium on nutritional significance of selenium (Factor 3). Federation Proceedings 20, 665 PP.

Somayajulu,B.L.K Meghani, C.A. and Borole, B.V. Uranium in drinking waters. Rejection in human body, PP.528-532 Management of Environment, ed. Patel, B. Wiley Eastern Limited, 578PP.

Sreeramamurthy K., (1989) Effect of irrigation water quality on soil-rice interrelationships. Annamalai Univ. Pub., Annamalainagar, 200 PP.

Stewart KM and Markello SJ., (1974) Seasonal Variations in concentrations of nitrate and total phosphorus and calculate nutrient loading for six lakes Western New York, Hydrobiologia, 44 (1), PP. 61-89.

Swedish Expert Group (1971) Methyl mercury in fish. A toxicologic-epedemiologic evaluation of risks. Nordisk hygienisk Tidskrift, Suppl.4.

Szabolcs I and Darab C., (1964) The influence of irrigation water of high sodium carbonate content of soils. proc. 8th Intern. Congress of Isss, Trans, II, PP. 803-812.

Taylor EWE., (1960) The pollution of surface and underground waters. British Water Works Association Journal, 42, PP. 582-603.

Towill LE et al., (1978) Review of the environmental effects of pollutants, III chromium, Cincinnati, US Dept. of Commerce, National Technical Information Service (PB-282-796).

Underwood, EJ., (1977) Trace elements in human and animal nutrition. Academic Press, New York.

U.S. Environmental Protection Agency (1976) Quality criteria for water, EPA 440/9-76-023, Washington D.

U.S. Enviornmental protection Agency (1977) Process design manual for land treatment of municipal wastewater, Rep.625/1-77- 008,chapter.2.

U.S. Environmental Protection Agency (1979) Acid rain, office of Research and Development Summary Rep. 600/8-79-028, PP. 3-6.

U.S. Salinity Laborartory (1954) Diagnosis and improvement of saline and alkali soils. US Dept. Agri., Agric.Handbook, 60, 160 pp.

Vallei BL. et al., (1960) Arsenic toxicology and biochemistry. A.M.A archives of industrial health, 21, 132 pp.

Watt BK and Merrill A.L., (1963) Composition of foods - raw, processed, prepared. Washington DC, Revised USDA Agriculture handbook,8.

Welby C.W., (1984) Groundwater yields and inventory for land-use planning in crystalline rock areas of wake country North Carolina. Water Res. Bull, 20, PP. 875-882.

Wilcox L.V., (1955) Classification and use of irrigation waters. U.S.Dept. Agri. Circ. 969.

World Health Organization (1970) Fluoride and human health (Monograph Serices No. 59).

World Health Organization (1972) Technical Report Series, 505, Evaluation of certain food additives and the contaminants; mercury, lead and cadmium.

World Health Organization (1973) Technical Report Series, 532, Trace elements in human nutrition.

World Health Organization (1977) Lead, Geneva Environmental Health Criteria 3.

World Health Organization (1978) Nitrates, nitrites and N-nitroso compounds. Geneva, Environmental Health Criteria 5.

World Health Organization (1984) Guidelines for drinking water quality-II 333pp.

Xiaoshu Chem et al., (1980) The relations of selenium and keshan disease. Biological trace element research, 2, p.91.

6

SALTWATER INTRUSION

D. CHANDRASEKHARAM

Department of Earth Sciences,
Indian Institute of Technology, Bombay 400 076

Many groundwater bodies often get mixed with saline water. Seawater is the most common source affecting the quality of groundwater in coastal aquifers in many areas. Besides seawater, industrial and agricultural activities also make the groundwater saline. This chapter emphasizes the seawater intrusion mechanism along coastal aquifers with case studies on the top unconfined laterite aquifer along the coastal belt of Kerala.

SEAWATER INTRUSION MECHANISM

Along the coast seawater inrtusion into aquifers, that are in hydraulic continuity with sea, occurs either due to low groundwater flow rate, high tides and lowering of water table due to excess withdrawal of groundwater. If the water table fluctuation occurs above the mean sea level, seawater-fresh water interface also fluctuates. If the water table falls below the mean sea level, negative gradient sets in allowing seawater to flow into the aquifer.

Since the seawater has higher density (ρ=1.025 gm/cm^3) than fresh water (ρ=1.0 gm/cm^3) fresh water tends to float on the seawater. To evaluate the nature of seawater intrusion mechanism, it is necessary to examine the disposition of the seawater-freshwater interface along coastal aquifers. The relationship between seawater and fresh water was studied by Ghyben (1888) and Herzberg(1901) who proposed based on density factor, an equation that relates these two liquids. Their analysis assumes simple hydrostatic conditions in a homogenous unconfined coastal aquifer which is in hydraulic connection with the sea. The equation is given below and the parameters related to the equation are shown in figure 1.

GROUND WATER
Editor: P. Sankara Pitchaiah, © 1994 Scientific Publishers, Jodhpur

$$\rho_s g Z_s = \rho_f g (Z_s + Z_w)$$

$$Z_s = \frac{\rho_f}{\rho_s - \rho_f} Z_w$$

$$Z_s = 40 Z_w$$

Assuming that the water table slopes towards the coast and is at an angle to the sea level, the seawater interface dips at an angle less than 90° with reference to the sea level.

If the water table fluctuation is ΔZw then from the Ghyben-Herzberg equation $\Delta Zs=40$ Zw. This means, if the water table is lowered by 1 metre the interface rises by 40 metres.

The Ghyben-Herzberg relationship, under real conditions underestimates the depth to the interface, for, the hydrostatic assumption of the analysis does not satisfy where the fresh water flow into the sea takes place. A more realistic method to locate the exact position of the interface was suggested by Hubbert(1940) by projecting the equipotential lines on to the interface and water table. This method helps in determining the position of the interface to certain degree of confidence. Hubbert's model for steady state flow is shown in figure 2.

Both the above concepts assume the interface to be a sharp boundary and a smooth surface while under field conditions it is a zone of diffusion having a concentration gradient with the salinity increasing with depth. The relative salinity (Sr) can be calculated if the salinity (c) at a particular depth is known. Thus

$$Sr = 100 (c\text{–}Cf/Cs\text{–}Cf)$$

where Cs and Cf are salinity of seawater and fresh water respectively. Studies carried out on Biscayne aquifer in Florida (Cooper, 1959) and Lusczynski and Swarzenski, (1966) show that the interface assumes irregular shape depending on the aquifer material and properties. Similarly, it is difficult to draw the interface configuration in aquifers which are in hydraulic continuity with adjacent river near the river mouth. In such areas due to high tide, seawater fresh water mixture enters the aquifer and disturbs the interface zone.

Under field conditions it is possible to understand the mechanism of seawater intrusion using seasonal water table data on a network of observation wells and the chemistry of groundwater together with simple seawater-freshwater mixing chemical models.

Besides seawater, sediments of marine origin locked within continental marginal basins forming a part of the aquifer also makes the fresh water saline.

In the succeeding part of this chapter case histories are presented

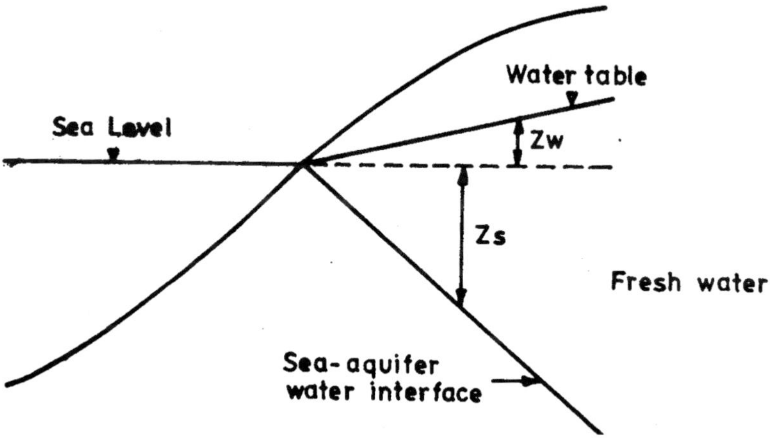

Fig. 1 : Seawater-freshwater interface in an unconfined coastal aquifer under hydrostatic condition.

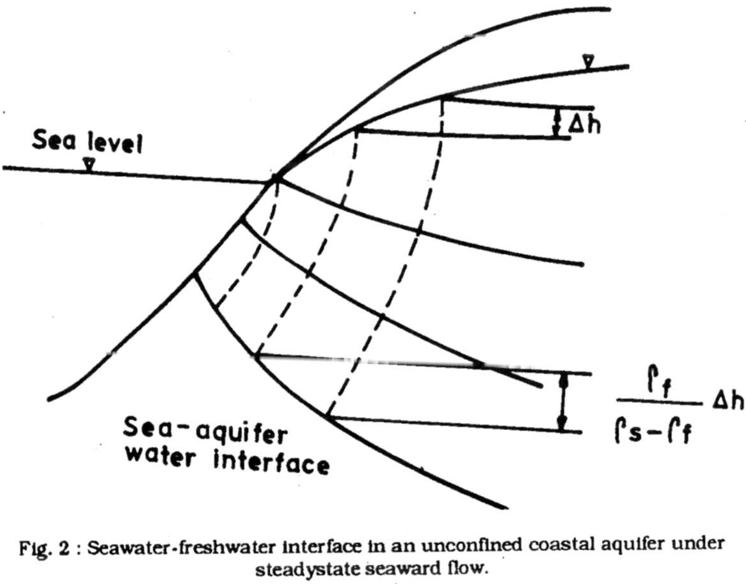

Fig. 2 : Seawater-freshwater interface in an unconfined coastal aquifer under steadystate seaward flow.

Table 1 : a : Depth to water table from the ground level in the wells

Well No.		Premonsoon			Monsoon			Postmonsoon		
		1980	1981	1982	1980	1981	1982	1980	1981	1982
MT1	(2.0)	1.4	2.0	1.6	1.8	1.8	1.5	1.2	1.5	1.8
MT2	(7.1)	5.7	5.3	5.4	–	2.3	0.2	4.0	4.2	4.9
MT3	(4.3)	3.8	3.5	3.7	1.4	1.2	1.0	2.0	2.4	2.9
MT4	(3.2)	2.6	2.8	2.7	2.4	2.2	2.3	2.5	2.6	2.7
MT7	(4.2)	3.0	3.2	–	2.7	2.8	–	3.6	3.5	–
MT15	(2.4)	1.8	2.0	–	0.8	2.3	–	1.6	1.5	–
B : Water levels with reference to mean sea level										
MT1		+0.6	0.0	+0.4	+0.2	+0.2	+0.5	+0.8	+0.5	+0.2
MT2		–0.7	–0.3	–0.4	–	+2.7	+4.8	+1.0	+0.8	+0.1
MT3		+1.2	+1.5	+1.3	+3.6	+3.8	+4.0	+3.0	+2.6	+2.1
MT4		–0.6	–0.8	–0.7	–0.4	–0.2	–0.3	–0.5	–0.6	–0.7
MT7		–1.0	–1.2	–	–0.7	–0.8	–	–1.6	–1.5	–
MT15		+0.2	0	–	+0.1	–0.3	–	+1.4	+1.5	–

All values in meters. Numbers in parentheses indicate total depth of the well from ground level.

based on studies carried out on the top unconfined laterite aquifer along the coastal belt of Kerala.

CASE STUDY I

Seawater intrusion along the coastal aquifer :

The top unconfined laterite aquifer along Malappuram district of Kerala is extensively tapped to meet the groundwater requirement of the area. Water is drawn through open dug wells whose density ranges from 200 to 300 in an area of 1 sq.km. The thickness of the aquifer varies from 3 to 15 metres. These wells partially penetrate the aquifer. Some of the wells show evidence of seawater mixing as indicated by the predominance of chloride ions over the bicarbonate ions in the water. In order to understand the mechanism of seawater intrusion in this region, an area of about 10 sq.km (Fig. 3) was selected for this study. Water levels in six permanent observation wells with depths varying from 2 to 7 metres were observed for three years (1980, 81 and 82) during premonsoon. monsoon and postmonsoon seasons. Water samples were collected during these periods and analysed for major ions concentration.

The water table levels in the six wells with reference to ground level and mean sea level are shown in table 1.

Table 2. Water levels in the open wells along Manjeswar-Kasargod coast (above m.s.l)

Well No.	1	2	3
K1	0.8	1.5	0.9
K2	2.7	1.3	0.4
K3	1.3	2.2	1.3
K4	4.3	5.2	4.5
K6	1.1	6.3	1.7
K7	1.4	2.6	1.7
K9	1.2	2.1	1.3
K10	0.4	0.8	0.6
K11	2.2	4.4	2.8
K13	1.4	3.3	1.7
K14	2.2	4.0	3.3
K15	3.2	3.5	3.1
K17	1.6	2.5	2.4
K18	12.0	13.2	12.4
K19	7.1	8.5	7.6
K20	1.7	1.8	1.9

All values in meters. 1, 2, and 3 : premonsoon, monsoon and postmonsoon data respectively.

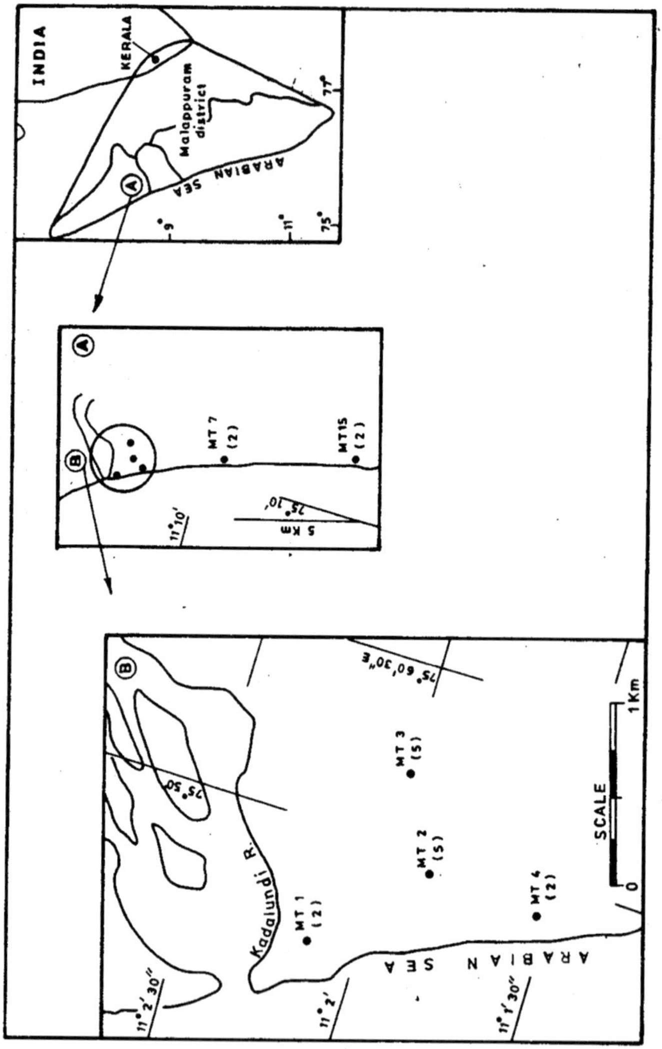

Fig. 3 : Map showing the area of investigation along Malappuram coast and the location of observation wells. Numbers in paranthesis indicate elevation of the ground at that point in meters.

The water table in well MT3 which is away from the coast indicates that the water table gradient is towards the sea but the gradient at certain locations is reversed permanently (MT4, MT7) and is seasonal in certain other locations (MT2 and MT15).

A detailed account on the chemical characteristics of the groundwater of this region was given by Chandrasekharam (1983, 1993). Groundwater occurring in laterite aquifer is of bicarbonate type (Fig. 4) but wells showing reverse gradient have chloride ions in excess of bicarbonate ions. The Cl, SO4 and HCO3 ions can be used to calssify the water into bicarbonate group and chloride group (Fig. 4). The wells belonging to chloride group can further be classified into non-saline (MT3), seasonally saline (MT3, MT15) and permanently saline (MT4 and MT7) wells. The Cl+SO4/HCO3 ratio of water in wells whose water table falls below mean sea level lie above unity thus establishing a relationship between water table and water chemistry (see Table 1 and Fig. 4). The water sample mixing with seawater fall on the freshwater-seawater mixing line (Fig. 4) indicating a simple mixing process between the two end members.

In order to understand the mechanism of seawater intrusion, it is necessary to quantify the amount of seawater in the saline wells. This is accomplished by using a simple two component mixing chemical model. This model developed earlier (Chandrasekharam, 1993) is shown in figure 5. The water from MT4, located near the coast is having about 1% seawater while MT2, located away from the coast is having 0.5% seawater. If Ghyben-Herzberg seawater intrusion mechanism is envisaged, then water in these shallow wells, whose water table falls below the mean sea level, should yield water whose NaCl concentration should nearly equal seawater. Thus seawater percent in MT7 should be much higher than MT4.

The water in MT1 will always be saline because of its proximity to the river mouth and its water quality is independent of the water table fluctuation.

This study demonstrates that seawater intrusion mechanism here is not related to Ghyben-Herzberg principle. The seasonal variation of NaCl concentration in the coastal aquifer can be explained by assuming a simple mixing process where, due to the prevailing negative gradient, dilute seawater present in the zone extending between the coast and observation wells enters the wells. Inflow of seawater due to reverse gradient is quite common in many coastal aquifers like those occurring along Long Island, New York and Miami, Florida USA (Todd, 1974). Overdraft is the main cause for lowering of the water table below the mean sea level in these regions.

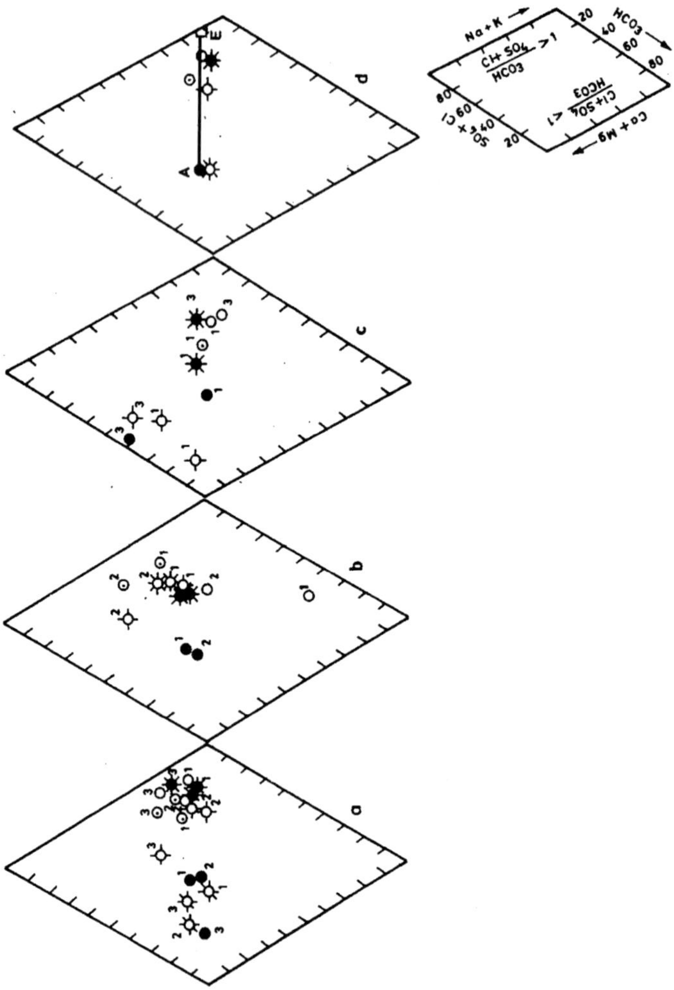

Fig. 4 : Piper diagram showing the geochemical facies of ground water along the Malappuram coast.
O WELL MT1; ◆ WELL MT2; ● WELL MT3; ✱ WELL MT4;
ʘ WELL MT7; ✲ WELL MT15; ▢ ARABIAN SEAWATER;
O 1, 2 and 3 : 1980, 1981 and 1982 data respectively.
a : Premonsoon; b : Monsoon; c : Postmonsoon plots.d ; Average plots of premonsoon data A-E : Seawater-aquiferwater tie line. Inset shows the scale and distribution of various facies.

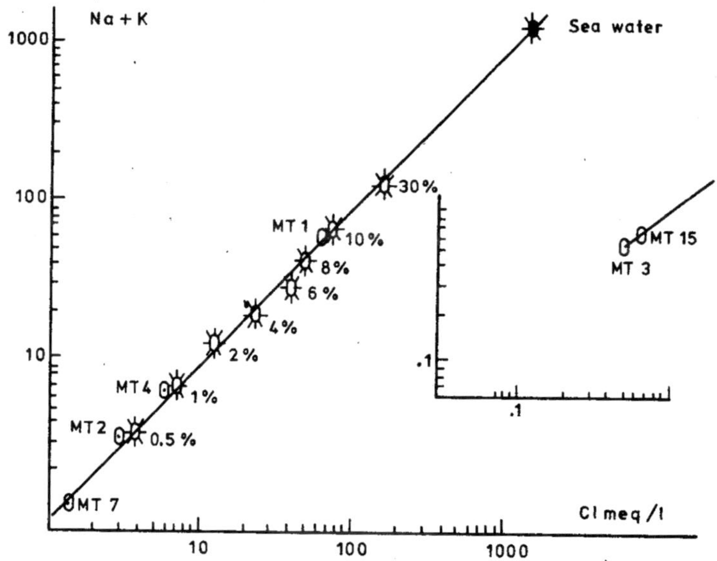

Fig. 5 : Aquifer-seawater mixing model. Percentages indicate amount of sea water in freshwater from well MT3.

CASE STUDY II

Occurrence of saline water along the coast :

This study demonstrates that seawater need not be the source of saltwater in the coastal aquifers. Investigation were carried out along the northern part of Kerala coast extending between Manjeswar and Kasargod (Fig. 6). Here, like in the previous case, laterite constitutes the unconfined aquifer and lie above a thick zone of Tertiary sediments of marine origin (Raha *et.al.*, 1983). The sediments include sandstones, varigated clays and carbonaceous clays. Pyrite is found in abundance in these sediments. Groundwater is extensively being utilised in this region and is extracted through partially penetrating open wells whose number varies from 200-300 per sq.km.

Water table levels, observed in sixteen observation wells during three seasons of 1982-83 hydrological year (Table. 2) lie above the mean sea level. Thus the groundwater flow is towards the sea during all seasons along the coast. Water table fluctuations are confined to zone above the mean sea level.

The major ions concentration in the groundwater are shown in Table 3. Seawater from the coast was also analysed during the course of this investigation and the analysis is shown in the table 4. Samples

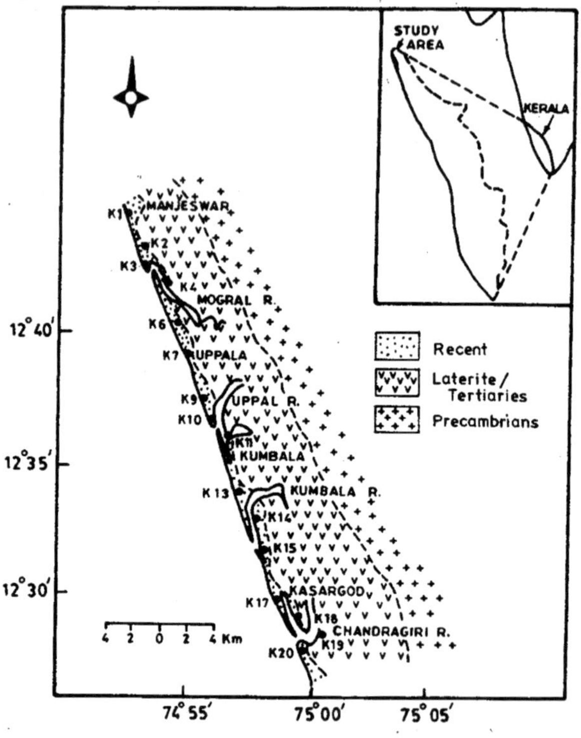

Fig. 6 : Geological and well location map of Manjeswar-Kasargod coast

from the river mouth (Chandragiri river, see fig. 6) and at several points away from the coast were also analysed (Table 4). Groundwater occurring in laterite are of HCO3 type and their HCO3/SO4+Cl ratio lie above unity (Chandrasekharam, 1989). Thus the groundwater occurring in this region should fall in the biocarbonate field when plotted on the Piper diagram (Piper, 1944). On the contrary, the well water samples record larger incidence of SO4 and Cl ions relative to HCO3 and fall in the SO4+Cl facies (Fig. 7). Although the samples show temporal variation, it is limited only to SO4+Cl facies (Fig. 7). A close observation of data presented in table 3 indicates lack of relationship between the Cl and Na+K ions. Sample, for example, with highest Cl concentration of 386 ppm registered only 40 ppm of total alkalies (K4, Table 3). The water samples record higher pH values (above 7) during monsoon while during the remaining seasons the pH values fall below 7 and a negative correlation is noticed between the pH and SO4 during postmonsoon season (Table 3).

Table 3 : Chemical analysis of water samples from wells along Manjeswar-Kasargod coast.

Well	No.	Ca	Mg	Na	K	HCO_3	SO_4	Cl	pH
K1	1	0.30	0.30	0.06	tr	0.12	0.17	0.36	6.1
	2	0.17	0.13	0.03	tr	0.10	0.10	0.15	7.7
	3	0.40	0.24	0.07	tr	0.27	0.17	0.31	6.1
K2	1	0.28	0.11	0.10	tr	0.07	0.02	0.41	5.5
	2	0.25	0.27	0.03	tr	0.19	0.01	0.35	7.9
	3	0.31	0.12	0.07	tr	0.21	0.02	0.30	6.1
K3	1	4.65	3.04	1.43	tr	0.62	1.43	7.09	6.5
	2	1.70	0.72	0.17	tr	0.52	1.00	1.01	8.0
	3	3.10	2.10	0.69	tr	0.46	0.67	4.67	nd
K4	1	7.81	3.81	1.76	tr	1.39	1.09	10.90	7.6
	2	2.65	1.16	0.08	tr	0.94	0.85	2.06	7.9
	3	1.60	4.20	0.22	tr	1.14	1.66	3.23	6.2
K6	1	0.18	0.50	0.03	tr	0.09	0.05	0.58	5.7
	2	0.30	0.30	0.02	tr	0.06	0.05	0.52	7.2
	3	0.48	0.37	0.03	tr	0.13	0.07	0.69	7.5
K7	1	0.95	0.16	0.03	tr	0.61	0.08	0.46	6.7
	2	0.21	0.16	0.04	tr	0.16	0.04	0.20	7.1
	3	0.53	0.14	0.07	tr	0.44	0.05	0.25	6.6
K9	1	0.75	0.45	0.06	tr	0.44	0.14	0.70	6.4
	2	0.68	0.26	0.04	tr	0.43	0.12	0.43	7.5
	3	0.75	0.26	0.08	tr	0.29	0.11	0.68	7.3
K10	1	7.60	3.14	0.39	tr	2.69	1.16	7.28	7.3
	2	1.61	0.57	0.14	tr	1.19	0.20	0.94	8.4
	3	4.00	1.81	0.35	0.02	2.62	1.36	2.18	7.2

Table contd...

Well	No.	Ca	Mg	Na	K	HCO_3	SO_4	Cl	pH
K11	1	0.42	0.10	0.03	tr	0.20	0.02	0.34	6.0
	2	0.77	0.23	0.02	tr	0.28	0.50	0.25	7.3
	3	0.44	0.76	0.03	tr	0.23	0.69	0.31	5.5
K13	1	0.98	0.14	0.04	tr	0.74	0.03	0.39	6.5
	2	0.82	0.16	0.05	tr	0.52	0.04	0.50	7.8
	3	0.50	0.32	0.07	tr	0.23	0.03	0.59	6.1
K14	1	0.65	0.44	0.06	tr	0.20	0.22	0.72	5.5
	2	0.66	0.45	0.15	tr	0.35	0.21	0.71	7.4
	3	0.37	0.41	0.18	tr	0.25	0.33	0.60	5.3
K15	1	1.65	0.60	0.07	tr	0.27	0.83	1.20	6.2
	2	0.66	0.46	0.05	tr	0.35	0.60	0.22	7.5
	3	1.05	0.96	0.05	tr	0.34	1.09	0.63	5.3
K17	1	5.20	2.92	0.27	tr	2.87	0.83	4.69	6.9
	2	2.00	0.99	0.13	tr	1.01	1.00	1.11	7.9
	3	2.34	1.46	0.14	tr	1.13	1.60	1.22	6.1
K19	1	1.11	0.71	0.05	tr	1.29	0.25	0.32	7.6
	2	1.81	1.57	0.47	tr	1.46	0.72	1.69	8.1
	3	1.76	1.70	0.67	tr	1.29	0.44	2.40	6.1
K20	1	0.90	0.48	0.16	tr	0.29	0.52	0.73	6.2
	2	3.28	1.60	0.74	tr	1.70	1.20	2.72	8.0
	3	1.25	1.09	0.05	tr	0.21	1.58	0.60	5.3

All values in epm. tr : in traces; nd = not determined

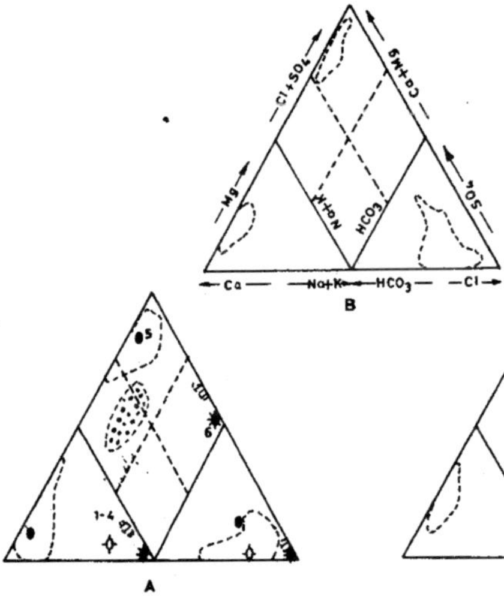

Fig. 7 : Piper diagram showing the various geochemical facies of groundwater along Manjeswar-Kasargod coast. A, B and C:
Fields of groundwater from the study area during premonsoon, monsoon and postmonsoon seasons respectively. Stippled field in A shows the field of groundwater in a typical laterite aquifer (Chnadrasekharam, 1989). Striped field shows the field of river- seawater mixture. For numbers refer Table 4.
a: Ca+Mg bicarbonate facies; b : Na+K bicarbonate facies; c: : Na+K chloride and sulphate facies; d : Ca+Mg chloride and sulphate facies.

Table 4. Chemical analyses of river and seawater

	1	2	3	4	5	6
pH	7.6	7.2	7.1	7.0	7.0	6.5
Ca	13.97	11.98	9.98	10.48	2.2	19.46
Mg	100.64	79.98	64.22	43.17	0.71	99.92
Na	316.70	273.91	200.00	106.52	0.22	1130.00
K	53.07	22.81	12.56	2.30	0.01	128.00
HCO_3	1.72	1.31	1.15	1.06	0.74	1.83
SO_4	41.37	34.73	29.12	20.92	0.98	75.47
Cl	441.17	352.62	256.50	140.42	1.41	1300.00

1 : sample near the river (Chandragiri river) mouth; 5 : sample about 15 km away from the mouth towards the upstreamside; 2-4 : samples at points inbetween locations 1 and 5; 6: Arabian seawater. Values in epm.

The higher incidence of SO4 and Cl ions relative to HCO3 is generally attributed to mixing of seawater, for, intrusion of seawater into the coastal aquifer is common. It has been demonstrated earlier (case study I), using sea-aquifer water mixing model, that intrusion of seawater into coastal aquifer occurs only when the gradient is reversed due to excess withdrawal of groundwater. However, in the present case, as evident from the water table (Table. 2) negative gradient is not observed in the wells and the Cl and Na+K ions do not show any systematic relationship which is otherwise expected if seawater mixing is taking place. Thus seawater intrusion is not the cause for higher incidence of Cl and SO4 ions in those wells. In order to examine the cause for higher incidence of SO4 and Cl ions, the chemical data were plotted on SO4/Cl vs SO4 and Mg/Ca vs Cl variation diagrams (Figs. 8,9). Water samples collected at different locations in the river away from the coast are also plotted in the above figures. The river water samples with different seawater concentration lie on the river-seawater tie line thus showing clear indication of mixing. By applying similar analogy, if the higher concentration of SO4 and Cl in the water samples is assumed to be due to seawater mixing, then the concentrations of Ca, Mg, SO4 and Cl ions in the mixtures should lie within the boundaries of the tie line between the Arabian seawater and groundwater field in figures 8 and 9. Further, if it is argued that the Ca and Mg concentrations of the mixtures need not lie between the two end members due to their sorption by the clays then at least the SO4 and Cl ions, which do not interfere with clay-water system, should lie

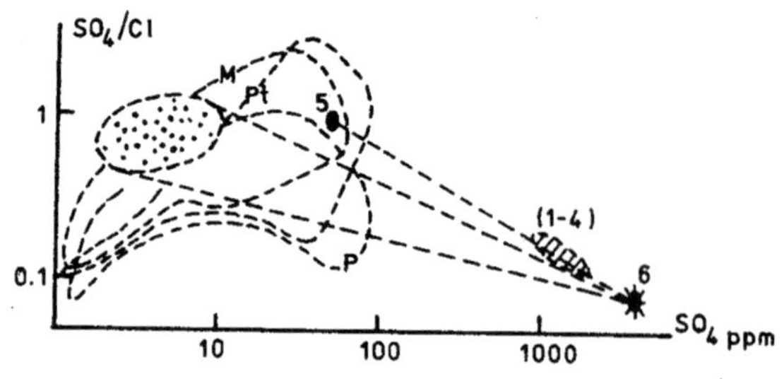

Fig. 8 : SO4/Cl vs SO4 diagram showing the fields of groundwater along Manjeswar-Kasargod coast. P: pre-monsoon; Pt : post-monsoon. M : monsoon; For symbols and numbers refer Figure 7.

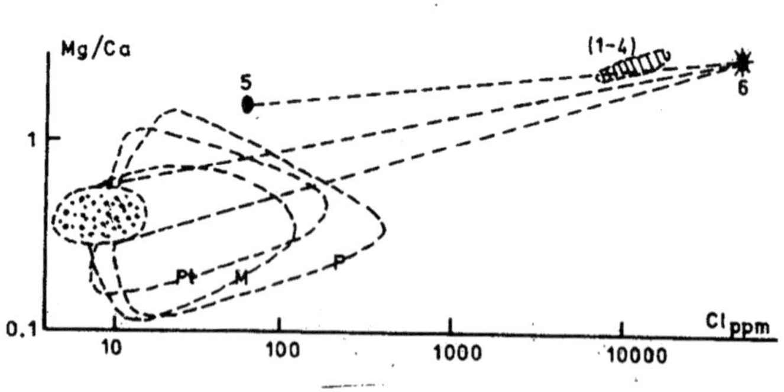

Fig. 9 : Mg/Ca vs Cl/diagram. For symbols and numbers refer Figure 7.

on this line.

The water table always lie above the sealevel during all the seasons of the year and the fluctuation in the water table is minimum in well K14 (1.1 m) and maximum in well K2 (2.3 m). If the Ghyben-Herzberg seawater intrusion mechanism is envisaged, then the SO4 and Cl concentration in these wells should show an increase relative to other wells. On the contrary, wells K3, K4 and K10 which show negligible water level fluctuation during the above seasons, record the highest SO4 and Cl concentrations (Table 3). Thus in this case the water level data and water chemistry put severe constraint in advocating seawater mixing process in this region. The alternate source which could supply SO4 and Cl ions to the aqueous system is the Tertiary sediments consisting of various members described previously. It is only natural to expect them to retain old marine salts at the time of deposition and release them subsequently during their interaction with water recharging the aquifer. Cases of groundwater contamination with connate water of marine origin were reported from Gujrat coast (Desai *et.al.*, 1979). Besides the old sea salts, iron sulphides which are present in the sediments, also contribute SO4 to groundwater. This is especially so during the postmonsoon season where the SO4 increase is considerable (Table 3). An inverse relationship between SO4 and pH in the water samples during this season (Table 3) gives evidence of breakdown of sulphides to SO4 thus releasing H ions to the aqueous system.

These two case studies clearly demonstrate that the Ghyben-Herzberg seawater intrusion mechanism need not be invoked always to explain occurrence of saline water in coastal unconfined aquifers.

CONTROL OF SEA WATER INTRUSION

Several methods can be adopted to protect the coastal aquifers from seawater intrusion and maintain the groundwater quality. Certain common methods are described below

Controlling extraction pattern

This is an effective method to control inflow of seawater into the aquifers. Based on pump test analysis groundwater extraction should be optimised to avoid lowering of the water level below the mean sea level. As evident from the case study I, normal hydraulic gradient is established in certain wells during the monsoon period due to excess precipitation and recharge. Optimum extraction of groundwater from such wells would help in maintaining the water table above mean sea level and obtain potable water throughout the year. Extraction should be stopped from wells whose water level lie permanently below mean sea level.

Artificial recharge

Using surface spreading method, the water table level can be raised to establish normal hydraulic gradient where the gradient is permanently reversed. Thus the excess surface runoff can effectively be utilised to control inflow of seawater into the aquifer. This method is possible to implement in regions of high rain fall.

Fresh water ridge or extraction barrier

A fresh water ridge parallel to the coast can be created by pumping water through injection wells (Fig. 10) so that the seawater freshwater interface can be kept in position away from the inland region and the water table can be lifted above mean sea level. Similarly a barrier can be created by extracting water through a line of wells parallel to the coast (Fig. 11). Seawater flows into the trough created by the extraction barrier and freshwater also flows into the barrier. Extraction along the barrier should be greater than that in the open wells towards inland region.

In all the above methods the recharge rate must be equal to the sum of the freshwater wasted to the sea necessary to keep the interface

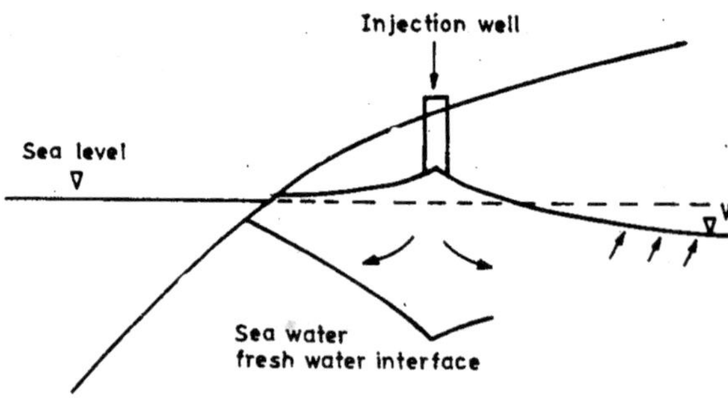

Fig. 10 : Creation of freshwater ridge through injection wells to prevent seawater intrusion.

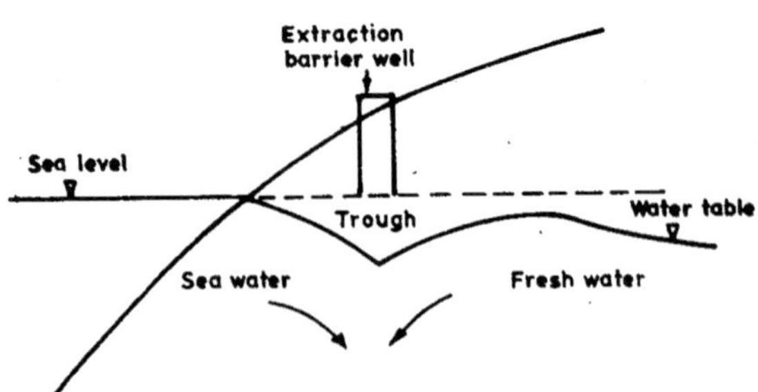

Fig. 11 : Creation of barrier trough through extraction wells to prevent seawater intrusion.

in position and the overdraft of the aquifer.

Construction of subsurface barrier.

The subsurface barrier reduces the permeability of an aquifer and prevents inflow of seawater into the aquifer (Fig. 12). This barrier also prevents wastage of potable water and allows large scale extraction of groundwater.

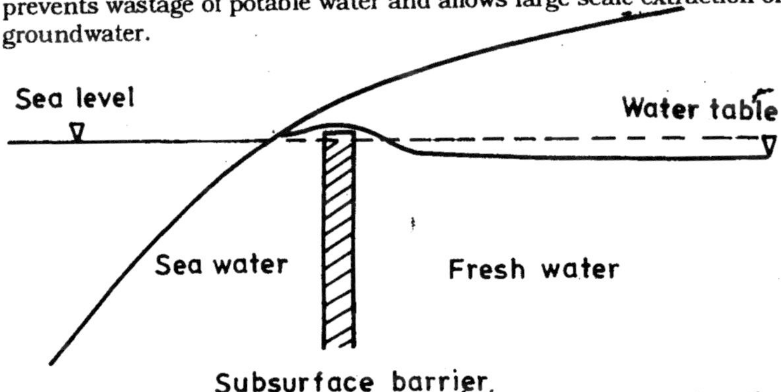

Fig. 12 : Subsurface barrier to prevent seawater intrusion into coastal aquifer.

Thus the seawater intrusion mechanism along the coastal aquifers is area specific and detailed investigation on the hydrological and hydrogeological characteristics of the aquifer and chemical characteristics of the groundwater should be carried out before seeking solutions to prevent inflow of seawater into the aquifer. In areas where geological strata are responsible for producing saline water, investigation prior to extraction of water should be carried out. Such sites should be avoided for groundwater extraction.

REFERENCES

Chandrasekharam, D. 1983 Saltwater intrusion mechanism along coastal aquifer, Malappuram district, Kerala. Research Report GW/R80/83, CWRDM, Calicut (unpub).

Chandrasekharam, D. 1989 Anomalous SO4-Cl groundwater in the coastal aquifer, Kerala. *Proceed. Indian Acad. Sci.* (Earth. Planet. Lett), 98, pp. 287-295.

Chandrasekharam, D 1993 Saltwater intrusion mechanism along coastal aquifer, Malappuram, Kerala in 'Groundwater' ed. M.K. Durga Prasad & P. Sankar Pichiah, Ajanta Book International (in press)

Cooper, H.H. 1959 A hypothesis concerning the dynamic balance of freshwater and salt water in a coastal aquifer. *J. Geophy. Res.*, 64 (4), pp. 461-467.

Desai, B.I, Gupta, S.K., Shah, M.V. and Sharma, S.C. 1979 Hydrochemical evidence of seawater intrusion along Mongrol - Chorwad coast of Saurashtra, Gujrat. *Hydrol. Sci. Bull.*, 24, pp. 71-82.

Ghyben B.W. 1888 *Nota in verband met de Vergenomen putboring nobij Amsterdam.* Tidsch. Von het Koninklijk Institut van Ingenieurs. The Hague.

Netherlands, pp. 8-22.

Herzberg, B. 1901 Die wasserversorgung einger Nordseebader. J. *Gashe. Wasser.*, 44 pp. 815-819.

Hubbert. M.K. 1940 The theroy of groundwater motion, J. Geol., 48 pp. 785-944.

Lusczynski, N.J. and Swarzenski, W.V. 1966 Saltwater encroachment in southern Nassau and southeastern Queens counties, Long Island, New York. *Geol. Sur. Am. Wat*-SUp. Paper 1613F, 75p.

Piper, A.M. 1944 A graphic procedure in the geochemical interpretation of water ananlysis. *Am. Geophy. Union, Trans.*, 25, pp. 914-923.

Raha, P.K., Sinha Roy, S. and Rajendran, C.P. 1983 A new approach to the lithostratigraphy of the Cenozoic sequence of Kerala. *J. Geol. Soc. India*, 24, pp. 325-342.

Todd, D.K. 1974 Salt water intrusion and its control. *J. Am. Water Work. Ass.*, 66(3), pp. 180-186.

7

REMOTE SENSING FOR GROUNDWATER EXPLORATION

S. M. RAMASAMY

School of Earth Sciences
Bharathidasan University
Tiruchirapalli-620 023

INTRODUCTION

The art of groundwater inventory and exploration has ever remained a matter of great enigma to the scientists and the technologists owing to the heterogenity of the aquifer systems. Targetting the locales of groundwater in crystalline domains has been still more problematic, because of the wide range of rock types and the frequently differing nature of structural features.

In the olden days, the people have mainly resorted to groundwater targetting through traditional methods of water divining like dowsing, floral indicators, organic landforms like termites, etc. Subsequently when the science was born and grown they started locating the groundwater pools, using modern scientific methods by drilling and borehole analyses, pumptests and aquifer characteristics studies, geophysical surveys and geochemical appraisals, etc. Even after such scientific techniques, the success limit was greatly on the lower side. The advent of modern remote sensing technology has totally changed the ethics of groundwater exploration as its synoptivity has aided the precise demarcation of the aquifer systems and on the basis which the hydrogeologists have started precisely pin-pointing the aquifers.

The present paper out the scintillating input that is provided by the remotely sensed data in groundwater exploration. In addition to the sequential presentation of details on the manifestation of various aquifer systems, appropriate case studies have also been cited from the author's own data bank.

GROUND WATER
Editor: P. Sankara Pitchaiah, © 1994 Scientific Publishers, Jodhpur

REMOTE SENSING FOR GROUNDWATER EXPLORATION

Control of Groundwater

Groundwater is basically controlled by lithology, structures (folds, faults, fractures and joints) and geomorphic landforms (denudational, fluvial, marine and aeolian).

Mapping of lithologically controlled aquifers

The primary porosity of the rocks is one of the fundamental parameters which controls the groundwater mobility in the aquifer systems. Hence, groundwater yield greatly differs from sedimentary to an igneous, and to the metamorphic rocks. Such differences in their porosity can be mapped precisely by certain photocharacters like tone and texture from the satellite imagery or from aerial photographs. The rocks with the high porosities invariably show light grey tone and coarser texture when compared to the rocks of low porosity which will normally exhibit dark grey tone and fine texture. Such tonal and textural grade maps can be prepared for an area either through the analysis of black and white panchromatic aerial photographs or by analysing the digitally enhanced satellite imageries.

The digital image processing of the Thematic Mapper satellite data for a part of Ariyalur region, Tamil Nadu clearly shows that the rocks like sandstones and limestones show light grey tone and coarser texture and the clays exhibit dark grey tone and fine texture. The charnockites and the gneisses on the contrary show medium to moderate tone and moderate texture. Hence from such satellite imagery, the regions of light grey tones and coarser textures can be mapped as favourable areas for groundwater and the regions of darker tone and fine texture can be negated as unfavourable areas for groundwater targetting. (Fig. 1.)

Still more precise information can also be brought out on the porosity and the permeability of the rock types by suitably enhancing the satellite pictures and developing computer aided models. As stated earlier porous and pervious lithologies show contrary texture on the contrary, impervious lithol show fine textural. Such textural contrasts of the rocks can be brought out from the satellite pictures by applying filters like sobel and Laplatian filters to the satellite multispectral photographs. Amongst different type of data the normalibed ratioed pictures generated by using the combination bands of 1+5/1-5 of Thematic Mapper data when filtered using Sobel 7 filter seems to give excellent information on the textural differences of the rocks which in turn give the porosity differences. In such filtered images the porous lithology shows tine coarser texture and impervious lithology shows fine texture (Fig. 2). Such filtered data of the satellite pictures provide

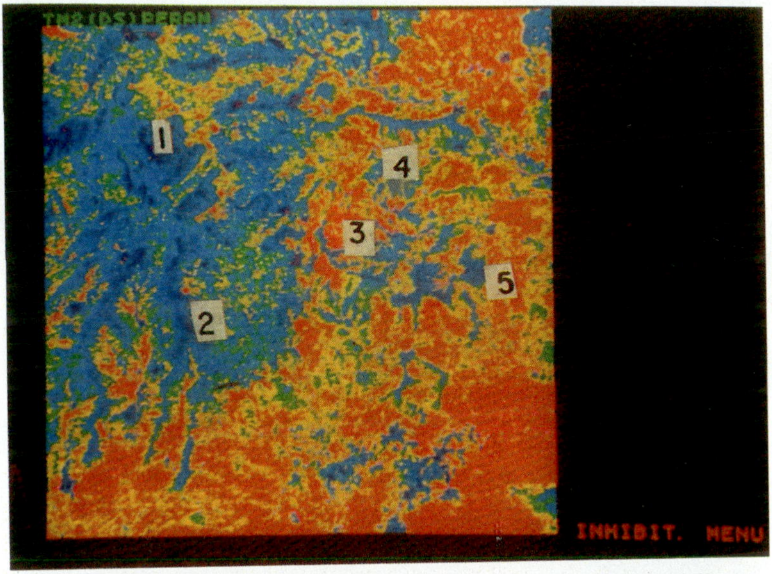

Fig 1. Satellite imagery of Perambalur area, Tiruchirapalli showing different porous, semi porous and impervious Lithologies. (**1**. Charnockite **2**. Genisses **3**. Upper Gondwana Sandstone **4**. Clay **5**. Other Sandstones)

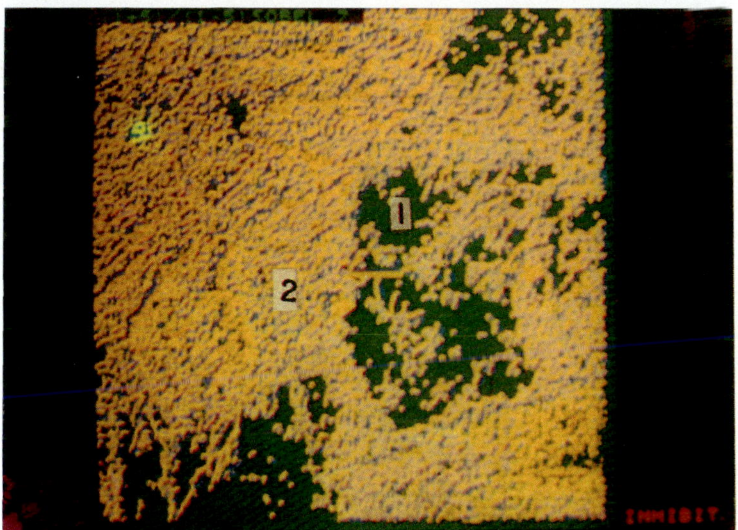

Fig 2. Digitally processed satellite imagery of Perambalur area. (1+5/1-5, Sobel-7 filtered and density sliced data.) **1**. Impervious clayey rocks. **2**. Pervious Sandstone, Charnockites, Gneisses and Limestones.

an excellent base for regional targetting of groundwater in lithologically controlled aquifer systems. (Fig. 2)

Mapping of aquifers controlled by the folded structures

In sedimentary, meta sedimentary and the meta igneous domains, the folded structures greatly influence the groundwater mobility an stability. The remotely sensed data provides an excellent and unparalleled information as the folded structures impart, linear, curvelinear, arcuate, semi-circular, elliptical and circular shapes to the hills. Such structures when occur in the subsurface conditions, on the basis of the soil tone variations, vegetation banding and also on the basis of the pattern exhibited by the parcelling of the cultivated fields, the structural trends can be brought out. After duly bringing out such structural trends whether for the hill domains or for the pediments or for the pediplains, on the basis of the drainage patterns, shadows and geomorphic expressions the subsurface configuration of the folds can be further brought out, (anticlines or synclines or domes or basins).For example, the radial drainages with centripetal pattern obviously suggest a basin and the radial centrifugal drainages on the contrary indicate a domal structure in the sub-surface.

After bringing out such fold styles, the regions of domal structures and sub-surface anticlines can be mapped out as groundwater barren tracts and on the contrary the regions of basinal structures and synclinal domains can be identified as favourable areas for groundwater. (Fig. 3)

One such a study was conducted for a part of typical crystalline aquifer systems in Tirupatur region of North Arcot district, Tamil Nadu (Usha, *et.al.*, 1989). The basement structures were brought out by the Thematic Mapper data on the basis of the soil tones and the vegetation tones. The water table data when correlated with such subsurface folded structures it was interestingly observed that the flow of groundwater was observed from the regions of subsurface anticlines to the regions of sub-surface synclines. (Fig. 4).

Mapping of fracture controlled aquifers

In crystalline aquifer systems, fractures play a dominant role in groundwater movement. Fractures are one such features that could be easily mappable in aerial and satellite imagery owing to the synoptic coverage of the earth surface by these platforms. The fractures do exhibit linear and curvelinear tonal patches, soil tone lineaments, straightness in vegetation, straight river coarses, long and linear hill ranges, etc. After mapping such fractures, these fractures can be tectonically classified as extension fractures (wide open in nature), shear fractures (normally tight fractures) and release fractures (marginally openup shallow fractures) by duly correlating the azimathal frequencies and the morphological expressions of these fractures with

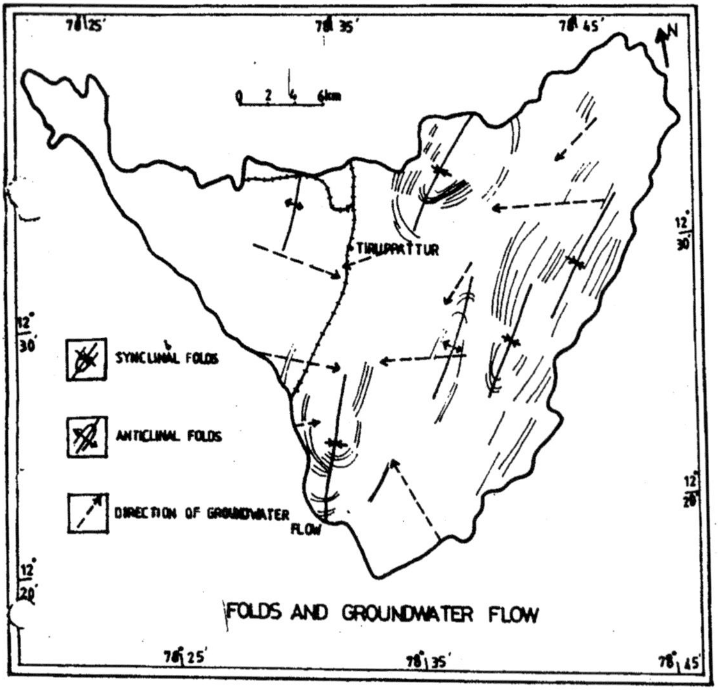

Fig. 4. After Usha, Ramasamy and Subramanian (1989)

folded structures. For example, normally, the extension fractures fall perpendicular to the axes of the folds, the release fractures are aligned parallel to the fold axes and the shear fractures form conjugate pattern with the fold axes. After classifying these lineaments, they can be correlated with groundwater flow and the contribution of different fracture systems can be brought out.

This could be done by developing iso-fracture lines separately for the extension, shear and release fractures and compare the linearity maxima of such contours with linearity maxima of the contours developed for the water table. Once the water bearing and water barren lineaments are identified then the groundwater locales can be pin-pointed accordingly.

Such models were developed for identifying the controls of folded structures and lineaments/fractures for entire south India (Ramasamy *et.al.*, 1989) carried out imagery interpretation and brought out the regional fold axes and prepared fractured density diagrams using remotely sensed data. Subsequently they have prepared groundwater level contours, groundawater yield contours and Transmissivity contours. Further, they have correlated the orientation of the fold axes and

Fig. 3. Satellite imagery of Kadavur area, Tiruchirapalli, Tamil Nadu showing exposed and burried geological structures. **1.** Basinal structures. **2.** Subsurface domal structure.

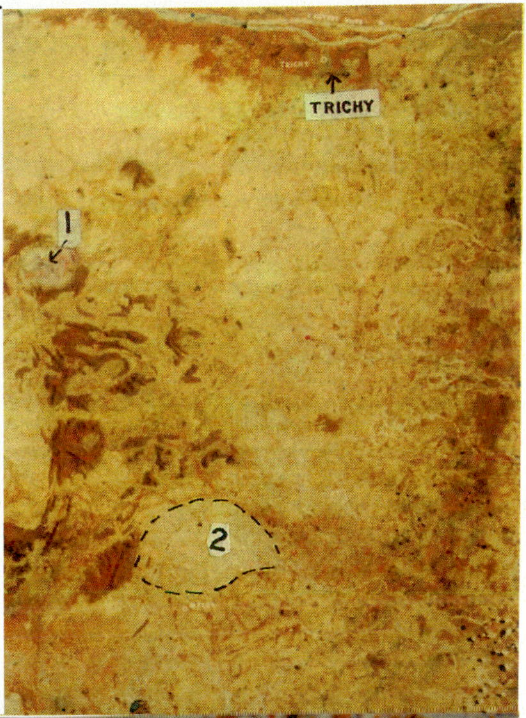

Fig 7. Satellite imagery of part of central Rajasthan showing groundwater bearing features. *A. Geographic locations* **1.** Bhilwara **2.** Banas river **3.** Kothari river *B. Denudational landforms* **4.** Denudational hills **5.** Pediments *C. Fluvial landforms* **6.** Flood plain **7.** Meander scars **8.** Meander loop

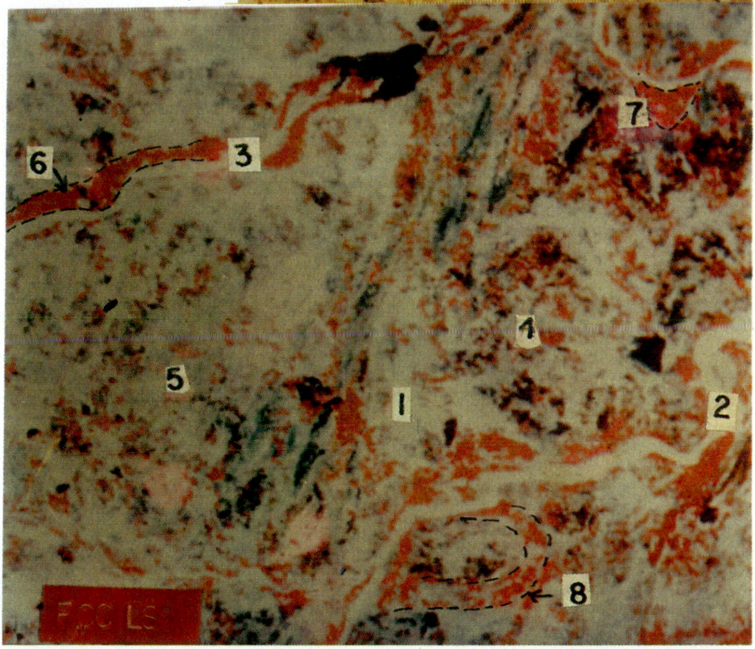

the linearity maxima of the fracture densities with groundwater yield linearity maxima and T-value contour maxima. On the basis of such an integrated study they have demonstrated that the groundwater is controlled by the folded structures in southern Tamil Nadu and controlled by the line-aments and fractures in the northern Tamil

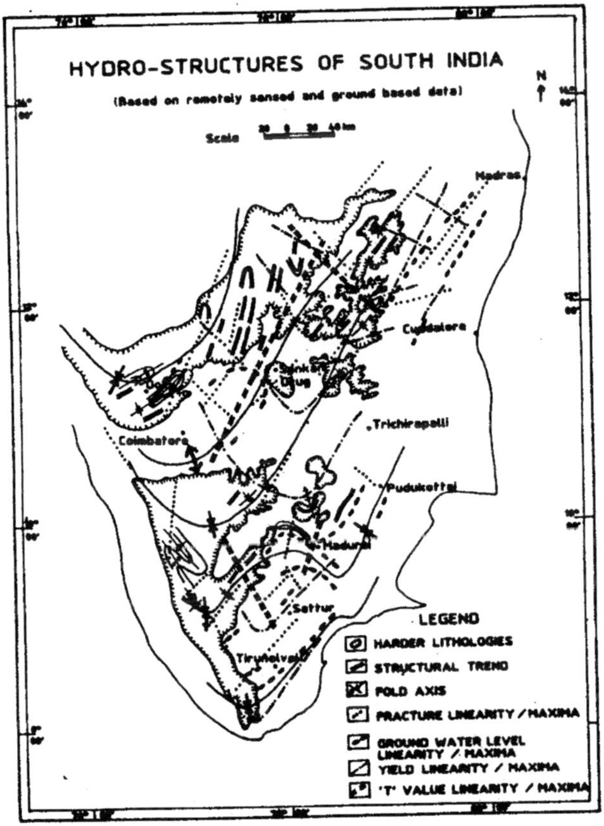

Fig 5. After Ramasamy, Balasubramanian, Thillaigovinda Rajan (1989)

Nadu. (Fig. 5)

Usha *et.al.*, (1989) have carried out detailed studies by preparing iso-fracture maps separately for the extension fractures, shear fractures, and the release fractures for Tirupat-ur region of North Arcot district (Tamil Nadu) and correlated the linearity maxima of such contours with groundwater flow and observed that the extension fractures are acting as best conduits for groundwater flow, and the release fractures come next. The shear fractures do not have any relation with groundwater flow. (Fig. 6.)

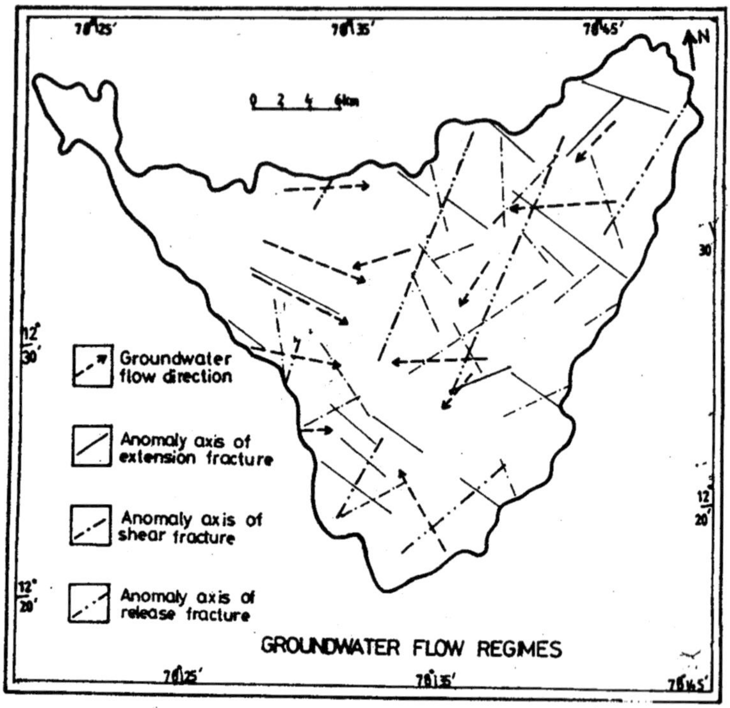

Fig. 6. After Usha, Ramasamy and Subramanian (1989)

Thus it is necessary to study the fracture systems in detail, in groundwater exploration after studying the lithological frame work. This vital information could be given only by remote sensing.

Mapping of aquifers controlled by denudational landforms

Once again in hard work rock aquifer systems, the weathered mantle acts as a potential reservoir for groundwater and in crystalline domains, the process of weathering causes denudational hills, debris slopes, talus cones, mud flows, pediments and pediplains. And amongst these, the talus materials, pediments and the pediplains form favourable groundwater locales. Such geomorphic units express characteristic and vivid signatures in remotely sensed data in the form of varying tonal grades, textural contrasts, vegetative covers, landuse practices etc. The processed satellite pictures like contrast stretched color composites give excellent information of such landforms.

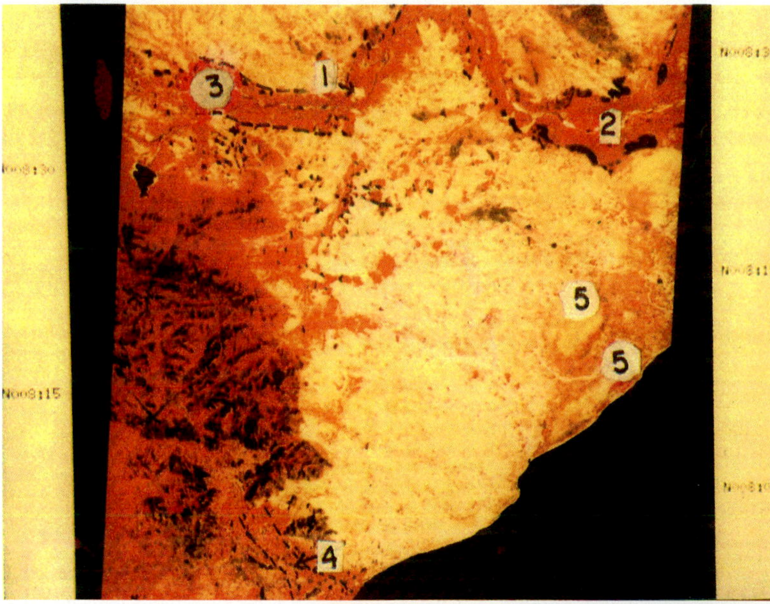

Fig. 8A. IRS–IA (LISS II) Satellite imagery of Tirunelveli region showing different groundwater bearing features. *A. Geographic locations* **1.** Tirunelveli **2.** Tamraparani river *B. Groundwater bearing features* **3.** Food plains **4.** Burried channels **5.** Beach dunes and Beach ridges

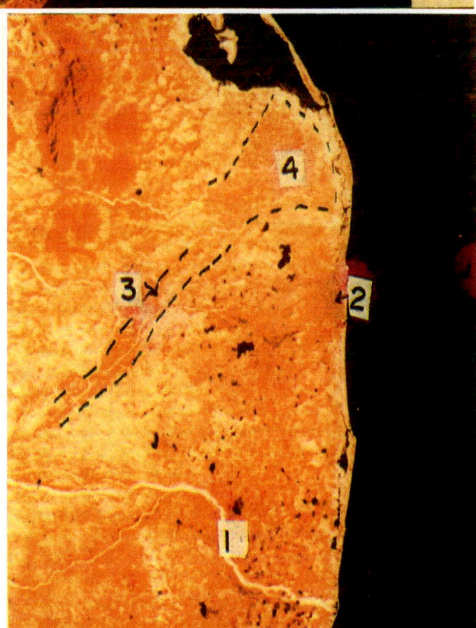

Fig. 8B. Satellite imagery of Palar-Madras region showing groundwater bearing features. *A. Geographic locations* **1.** Palar **2.** Madras city *B. water bearing features* **3.** Old course of Palar **4.** Palaeodelta

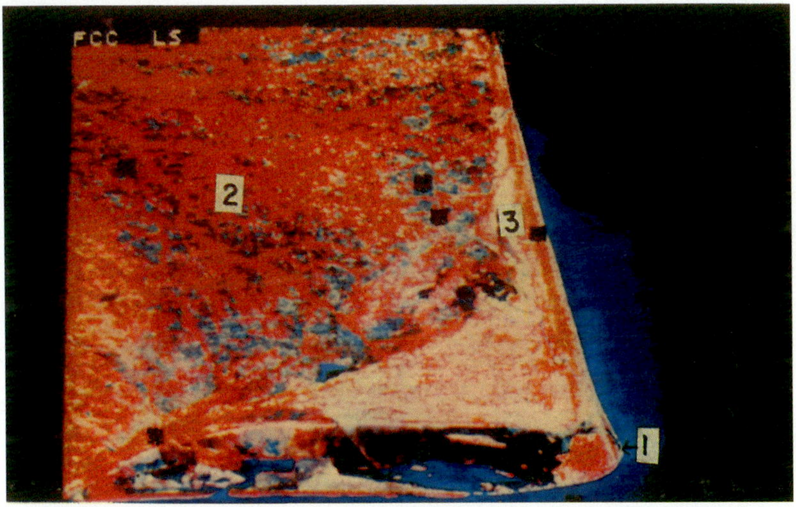

Fig. 9. Satellite imagery of point Calimere region of Tamil Nadu showing groundwater bearing fluvial and marine geomorphic features. *A. Geographic locations* **1.** Point Calimere *B. Water bearing features* **2.** Flood plains of

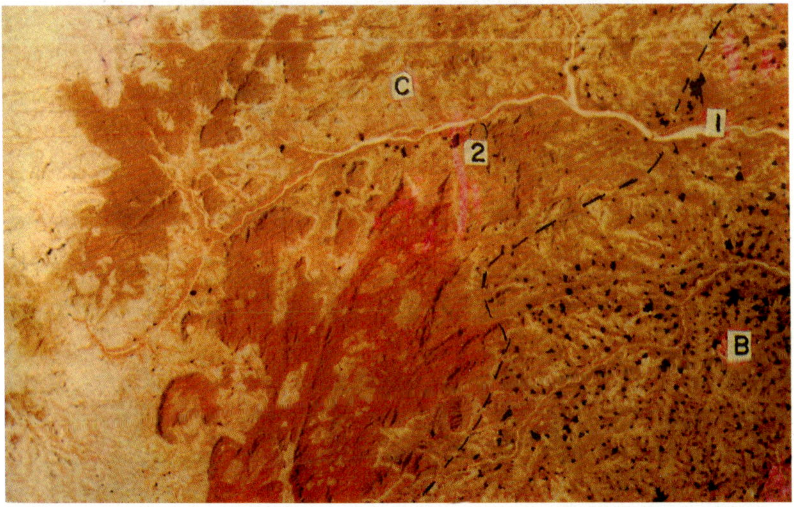

Fig 10. Satellite imagery of Palar–Cheyyar region showing contrast in the degree of weathering in rocks. *A. Geographic locations* **1.** Palar. **2.** Vellore *B. Zone of high concentration of tanks indicating high degree of weathering. C. Zone of low concentration of tanks indicating poor weathering.*

Such analyses carried out for the Banded Gneissic Complex of Rajasthan using contrast stretched false color composites have clearly brought out different denudational landforms. (Fig. 7.)

Mapping of aquifers controlled by fluvial landforms

It is an established fact that the river regimes are the ever yielding groundwater reservoirs. The rivers erode the materials in its youthful stage, oscillate here and there and develop flood plains, leveey and also migrate frequently in the matured stage and in the coastal plains travel with brownian movement, develope deltas and ultimately meet the sea. Features like flood plains, levees, buried channels, meander scars, oxbow lakes, etc., do act as groundwater reservoirs. These features can be precisely mapped out from black and white aerial photographs and satellite imagery. Water bearing regimes invariably appear in darker tone in black and white imagery and reddish color in color coded satellite pictures. (Figs. 7, 8A, 8B, 9).

Mapping of aquifers controlled by coastal landforms

The coastal region is a domain where the fluvial processes and the marine processes interact with each other. As a result, the sediments which were brought by river systems are subjected to further panning and sediment dispersion by the waves and finally deposited along the coast as long and linear sand bodies. In course of time when the sea recedes these sand bodies get stabilised as beach ridges. As these beach ridges are constituted by the coarse sands, they act as porous media for the accumulation of rain water and thus supply sweet and potable groundwater all along the coast. On the contrary the swales or the intervening depressions invariably bear saline water, hence these beach ridges must be mapped precisely along the coastal belt. In fact such beach ridges are occurring as bundles to a width of 5 to 20 kms all along the east coast of Tamil Nadu and bear good groundwater potential (Fig. No. 8A & 9).

Such beach ridges owing to their brighter tone, coarser texture and linear shapes can be mapped clearly from satellite and aerial remote sensing data.

Mapping of aquifers controlled by aeolian landforms

The aeolian landforms are basically characterised by longitudinal, transverse, barchan and parabolic dunes. Among these dunes, the barchan dunes are migratory dunes. Sand from these dunes whenever traverse the moisture rich terrains one attracted by such moistures and hence fall down as barchan dunes. Hence wherever such barchan dunes are found they mostly indicate moistured regimes which in turn is related to shallow groundwater conditions.

Such migratory dunes can be mapped easily using satellite imagery and also aerial photographs as they occur as cresent shaped bodies, form clusters and also express bright white signatures. Mapping of these dunes form an essential aspect for groundwater exploration in desert regions.

Mapping of aquifers on the basis of surface water bodies

The concentration, clustering and the frequency of surface water bodies (lakes and tanks) mostly indicate the highly weathered nature of the rocks. Hence this can be used as an and indirect indicator for judging the degree of weathering. Such water bodies can be mapped precisely from the satellite photographs taken under near infra- red ranges as electro magnetic radiation (EMR) is totally absorbed under such ranges. Hence by using such satellite data, the surface water bodies can be mapped and on the basis of their concentration, the degree of weathering can be indirectly assessed for further groundwater targetting. (Fig. 10)

Mapping of aquifer on the basis of soil moisture

At times the soil moisture too indicates the water table conditions. The regimes of higher soil moisture exhibits darker tone in black and white imagery and bluish tone in color coded satellite images. Thus on the basis of these satellite pictures the regimes of soil moisture can be mapped to delineate the groundwater potential zones.

REFERENCES

Ramasamy, SM., Thillai Govindarajan, S., and Balasubramainan, T. 1989 Remote Sensing based appropriate Methodology for the -groundwater exploration-A case study from Precambrians of South India. *Proceeding volume International workshop on Appropriate Methodologies for development and Management of groundwater resource in developing countries*, NGRI, Hyderabad, India, Oxford Press : pp. 341-344.

Usha, K., Ramasamy, SM., and Subramanian. SP., 1989 Fracture pattern modelling for groundwater targetting in hard rock terrain —A study aided by remote sensing technique. *Proceeding volume International Workshop on Appropriate Methodologies for development and management of groundwater resources in developing countries*. Hyderabad, India, Oxford Press : pp. 319-328.

8

GROUNDWATER EXPLORATION AND GEOPHYSICAL METHODS

P. N. BALLUKRAYA

Dept. of Applied Geology, University of Madras
Madras - 600 025

Exploration for groundwater has exercised man's mind ever since he started moving inland from perennial rivers in his quest for a better living. The need arose because, though groundwater is generally available in most places, the locations where it is present in adequate quantity are rather restricted. Till about the middle of nineteenth century, groundwater exploration was the sole preserve and prerogative of dowsers, preists, astrologers and even quacks, claiming divine powers. There were also a few rational thinkers in early historical times who made careful study of groundwater environment and used their observations in a scientific way for identifying potential groundwater bearing zones.

Modern day exploratory techniques were related to the development of earth sciences and the realization of the basic fact that the groundwater bearing material is a geological formation. Thus, study of geology is an essential part of groundwater exploration in that, the availability or otherwise of groundwater in an area is directly related to the type of rocks present there and their structural and stratigraphic configuration. The development of geophysical tools which help to "look" at what lies beneath the ground surface further revolutionized the exploration for natural resources. Thus, in the field of groundwater exploration too, geophysical methods find immense application and infact, a majority of present-day groundwater development projects have geophysical investigations invariably built into them.

In the application of geophysical methods for groundwater exploration, it is often misunderstood by many that they are used to directly detect groundwater. However, this is an erroneous assumption and the

GROUND WATER
Editor: P. Sankara Pitchaiah, © 1994 Scientific Publishers, Jodhpur

that geophysical tools help in locating geological formations/features/structures which may be favourable for the accumulation of groundwater therein and not groundwater *per se*. This is true of all geophysical explorations, wherein one looks for geophysical anomalies which are then interpreted in geological or geohydrological terms. Thus, while exploring for groundwater in an alluvial terrain, one aims at identifying a sand/sand stone formation at some depth from the geophysical data, as these sands generally contain groundwater rather than the clays which are normally the associated rock formations. On the other hand, in areas underlain by crystalline rock formations it is a thick zone of weathered/fractured rock which is the target of exploration.

GEOPHYSICAL METHODS

Physical Properties

Geophysical methods of exploration are based on certain physical properties of earth materials. These properties are measured and variations in their values in lateral or vertical directions are made use of for gathering subsurface information. The most important rock properties that are made use of are : density (gravity prospecting); magnetic susceptibility (magnetic prospecting); elasticity (seismic prospecting); electrical conductivity (electrical prospecting) and radioactivity (radiometric prospecting). Almost all these methods find their application in groundwater exploration, though it is the electrical methods which are by far the most widely used techniques in this work.

GRAVITY METHODS

This method makes use of the variations in the density of different minerals and therefore rock types, to map the subsurface from a series of measurements made on the ground surface or even from air using equipments known as Gravimeters. The property that is measured is the local acceleration of gravity (unit : gal or milligal). The variations in the measured gravity values, corrected for local altitudes and lattitudes reflect the lateral variations in the density of rocks extending to several tens of kilometers depth in that area. It is common to use the Bouguer gravity anomaly which is the observed gravity at a station + drift and tidal correction + combined elevation correction + terrain correction - lattitude correction.

Gravity method is relatively an inexpensive and fast exploratory technique particularly useful as a reconnaissance tool for large areas. Specifically, it may be helpful in determining the configuration of aquifers, locating areas of maximum sediment thickness, tracing axes of buried channels and in identifying buried bedrock-highs in areas underlain by alluvium and sediments.

Fig. 1 gives the Bouguer anomaly map over an alluvial area wherein

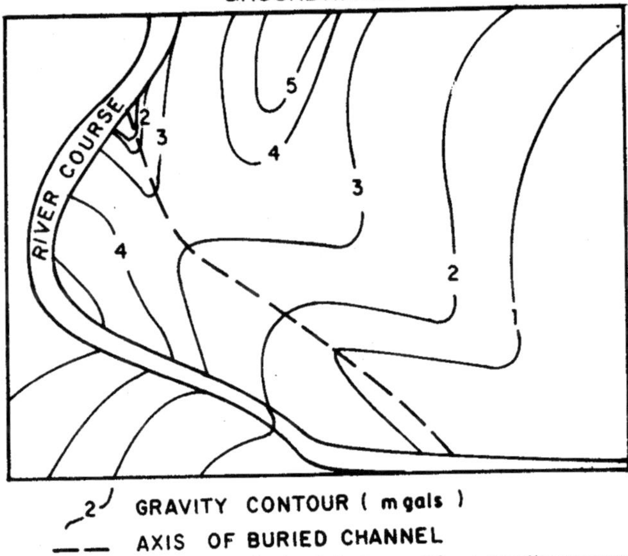

Fig. 1. Delineation of the axis of a buried channel from gravity measurements

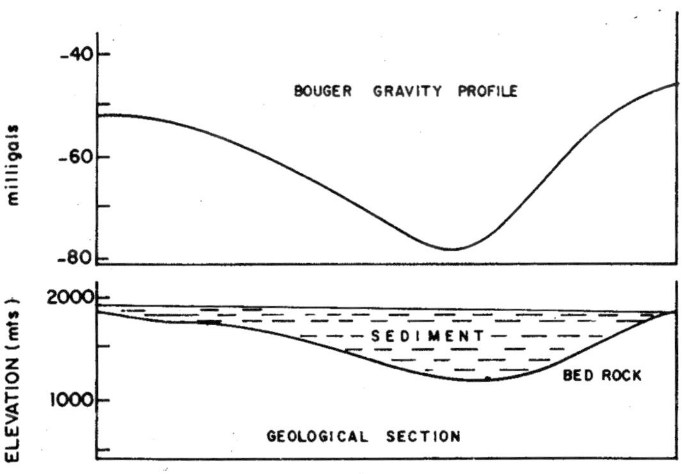

Fig. 2. Bouger gravity profile over a sedimentary basin

the axis of a buried river channel is clearly indicated. Fig. 2 shows the gravity anomaly to be expected over a deep sedimentary fill, with the lowest gravity values coinciding with the thickest column of sedimentary rocks. Highly sensitive gravimeters may even be used for estimating depth-to-water levels in an area. Fig. 3 shows the general relationship between ground waterlevels in an area and the nature of gravity anomalies obtained.

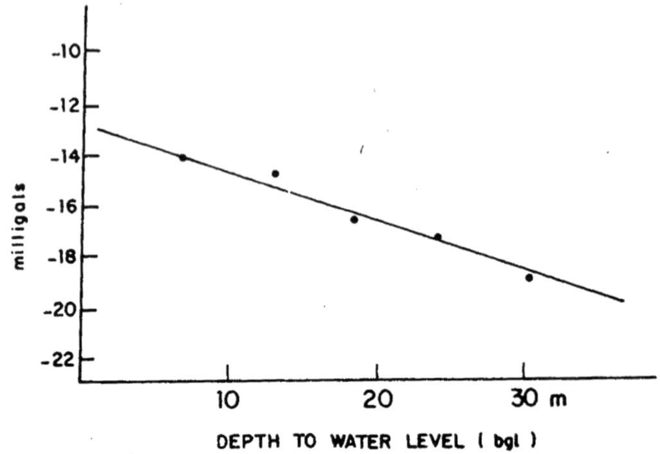

Fig. 3. Variation of measured gravity values with depth to water levels

MAGNETIC METHODS

The property of rocks to retain magnetism is made use of in magnetic prospecting techniques. The direction, intensity or gradient of the magnetic field in an area is measured and variations in these values are interpreted in geological terms. Either the total field or the vertical field may be measured for the purpose, using appropriate magnetometers.

The unit of magnetic intensity used in magnetic exploration is the gamma (or milligamma). The distortions in the earth's magnetic field, which ranges from about 25,000 gammas at the magnetic equator to about 69,000 gammas at the poles, are caused by the available magnetic materials in the earth's crust and mantle. The distortions in the magnetic field, caused by variations in the concentration of such magnetic minerals in an area, may be of the order of several tens or hundreds of milligamma, are made use of in magnetic prospecting. In groundwater exploration, the applicability of magnetic methods is rather limited and as is the case with gravity methods, is generally used as a reconnaissance tool. Two areas of applicability may be : (1) delineation of the basement configuration in a sedimentary basin and (2) mapping of basaltic aquifers that may be buried at a certain depth. Sedimentary rocks in general contain less magnetic minerals than the crystalline rocks and therefore, thicker the sedimentary column, greater will be the amplitude of the magnetic-low anomaly generated. (Fig. 4).

SEISMIC PROSPECTING

Seismic methods make use of the elasticity of rock materials. Artificially generated pulses of elastic energy are propagated through the ground and their arrival times at certain distances from the source

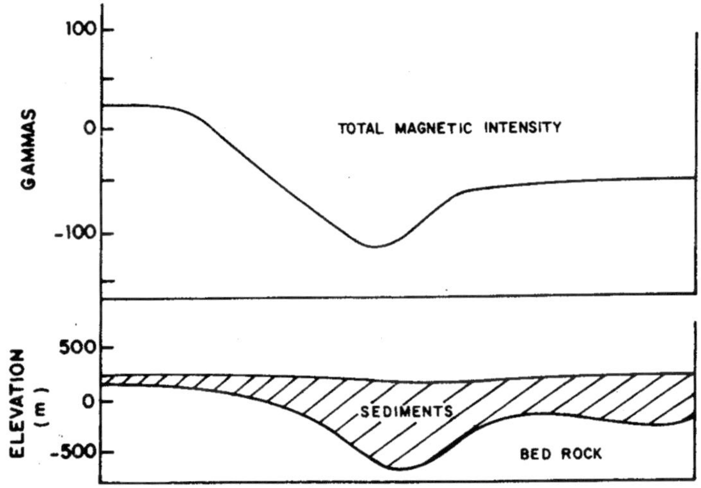

Fig. 4. Total magnetic intensity profile over a sedimentary basin

point are monitored by means of sensing devices known as geophones. Information regarding the subsurface is obtained from this data since the arrival-time is related to the velocity of energy propagation and the geometry of the path taken by the energy waves. Elastic energy is generated either by triggering explosives underground or using high-pressure hammers. Shallow investigations may even use a simple metal plate and a sledge-hammer to generate the waves.

The interpretation of seismic data can be quite complex, primarily due to the presence of large number of sub-surface layers of differing elastic properties. Thus, a wave may travel along the earth surface from the point of origin to the point at which it is picked up; it may get refracted at a layer boundary and this refracted energy may be picked up by the geophone; multiple refractions may occur by the presence of more number of sub-surface layers; the waves may be reflected from these various layer boundaries. All such waves will be picked up by a single geophone located at a point and therefore all these different signals have to be sifted to differentiate one another for arriving at an understanding of the sub-surface.

Fig. 5 shows a horizontally layered earth with a surface layer of velocity V_1 and the lower layer having a velocity V_2 with $V_1 < V_2$. Four paths of rays, along which the seismic waves may travel are shown and they are :

1. The direct ray, travelling horizontally from source to receiver (geophone);
2. totally reflected ray;

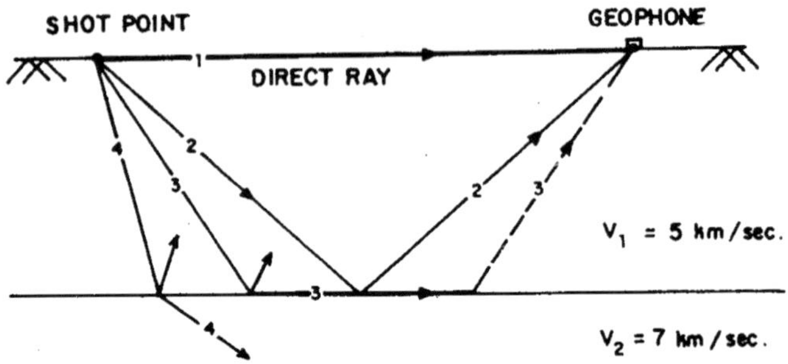

Fig. 5. Travel paths of seismic waves in layered earth

3. the ray, which is partly reflected and partly refracted and travelling along the layer-boundary at velocity, V_2 and

4. the ray which is partly reflected and partly refracted, travelling into the second medium.

The ray travelling along the interface at V_2 velocity generates a small pulse which eventually reaches the surface and the geophones located there. Exploration techniques which make use of the reflected energy are known as Reflection surveys while those using refracted energy are termed as Refraction surveys. Refraction surveys are more common in groundwater exploration, to determine the nature as well as number of sub-surface layers, their thickness and other characteristics.

Fig. 6 shows the waveform of compressional waves radiating from a source point in a two-layered earth and the arrival time versus distance graph. The geophones close to the shot point first receive the energy travelling along the surface. At greater distances, the refracted wave travelling at the velocity of the second layer (which is greater) reaches the surface first and therefore are picked up by the geophones as first arrivals. The energy travelling along the ground surface reaches these geophones slightly later as second arrivals. The plot of arrival times versus distances, therefore, will have two segments. The inverse of the segment slopes gives the respective travel velocities and the depth to the refracting layer, z, is given by (Nettleton, 1940) :

$$Z = \frac{d_c}{2} \sqrt{\frac{V_2 - V_1}{V_2 + V_1}} \quad \ldots\ldots (1)$$

where d_c is the critical distance (fig. 6). However, the above equation does not hold good where V_2 is less than V_1.

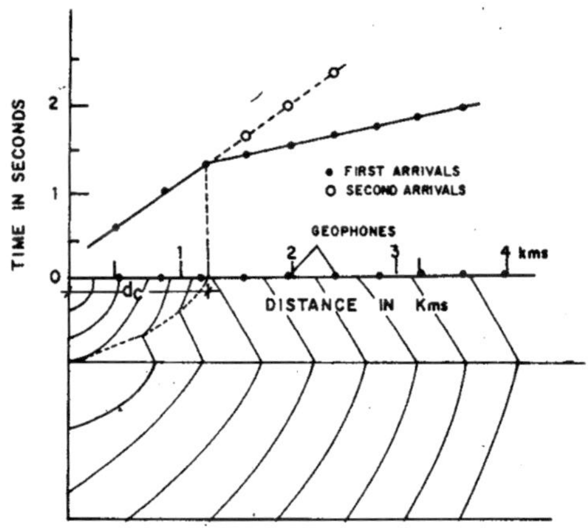

Fig. 6. Distance - travel time plot for a two layered earth and the wave phase pattern

In groundwater prospecting, seismic refraction investigations can be used to (1) determine thickness of sedimentary column in an area (2) determine the depth to bed rock (i.e. thickness of weathered rock) in areas underlain by crystalline rocks.

ELECTRICAL METHODS

General

One of the important physical properties of rocks and rock-forming minerals is their ability to conduct electric current. Electrical methods of prospecting make use of this property for inferring subsurface geology. Several electrical characteristics are employed for this purpose - electrical resistivity (or inversely, conductivity), spontaneous potential, induced polarization, telluric current etc. However, in groundwater exploration, it is the electrical resistivity method which is the most widely used technique.

ELECTRICAL RESISTIVITY METHOD

The electrical resistivity methods involve the measurement of the resistance, offered by the rock formations in an area, to the flow of an electric current of known intensity. The resistivity of a rock unit depends on its mineral composition and is influenced to a very large extent by the interstitial water content present therein.

Electrical resistivity method involves the measurement of surface potentials caused by the passage of an electric current, impressed on the ground from an artificial source. It is based on the validity of Ohm's law for linear conductors, $R = \Delta V/I$, where R is the resistance, in ohms, offered to the flow of current I (in amperes) and ΔV is the potential difference, in volts, across the two end faces of the conducting material. The resistance of a medium is directly proportional to its length L and inversely proportional to its cross-sectional area A, so that

$R \alpha L/A$

The electrical resistivity or the specific resistance, ρ of the conducting medium, then is

$$\rho = \frac{A}{L}R = \frac{A}{L} \cdot \frac{\Delta V}{I} \qquad \ldots\ldots (2)$$

Thus, the resistivity can be defined as the resistance offered by a material of unit dimensions and in the mks system its unit is a ohm-meter (Ωm).

The pioneering efforts of Conrad Schlumberger (1912-1914) are mainly responsible for the development of electrical resistivity method as an effective tool in geophysical exploration. In field measurements, current is introduced to the ground through two electrodes and the potential difference between another pair of electrodes is measured. The strength of the current impressed and the potential difference that is measured is then used to calculate the resistivity of the ground, taking into account the geometry of the electrode spread. Fig. 7 shows the arrangement and the current and potential lines in a homogeneous and isotropic medium. The potential difference between the two potential electrodes, M and N, due to the current introduced by the two current electrodes, A and B, is given by Fig. 7

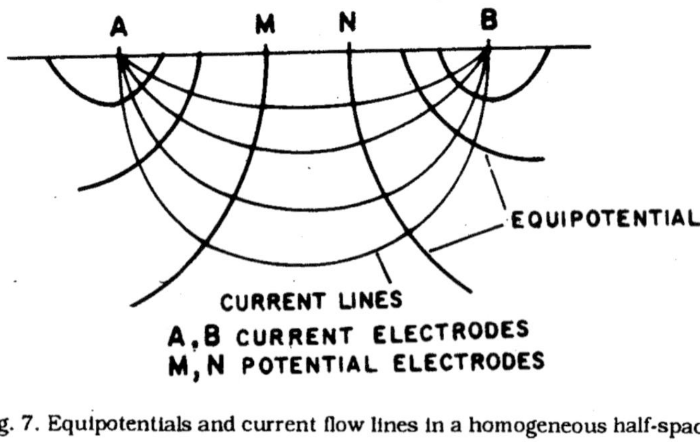

Fig. 7. Equipotentials and current flow lines in a homogeneous half-space

$$V_{M,N}^{A,B} = V_M^{A,B} - V_N^{A,B} = \frac{\rho I}{2\pi}\left(\frac{1}{AM} - \frac{1}{BM} - \frac{1}{AN} + \frac{1}{BN}\right)$$

where $V_M^{A,B}$ is the potential developed at M due to A and B and $V_N^{A,B}$ is the potential developed at N due to A and B; AM, AN, BM and BN are the respective inter-electrode distances.

The resistivity of the medium, therefore, is given by

$$\rho = \frac{2\pi}{\frac{1}{AM} - \frac{1}{BM} - \frac{1}{AN} + \frac{1}{BN}} \frac{\Delta V}{I} \qquad \ldots\ldots (3)$$

and the factor $2\pi \Big/ \left(\frac{1}{AM} - \frac{1}{BM} - \frac{1}{AN} + \frac{1}{BN}\right)$ is generally termed as the geometric constant of the electrode configuration used and is denoted by the letter K, so that

$$\rho = K \frac{\Delta V}{I}$$

The resistivity measured using the above equation is the true resistivity of the medium only when measurements are carried out over a homogeneous and isotropic material. However, for all practical purposes, the actual conditions in the field are far removed from this in that the earth's subsurface is made up of two or more layers of different rock units. Therefore, the value obtained from measurements over such a section is not the true resistivity, but only a "weighted" average of the materials in the subsurface and therefore is termed as the *apparent resistivity*. It may be noted here that while the true resistivity is specified for a particular material/rock formation, apparent resistivity values are for given electrode separations in an electrical resistivity measurement exercise. This apparent resistivity is the basic parameter in all resistivity investigations from which other parameters such as the true resistivities of the various rock units in a given area are determined. In the computation of the resistivity values using the four-electrode systems and equation 3, the current source is assumed to be a point source, which however is not the case in reality, as linear metal electrodes are used as current electrodes in the field. Therefore, the resistivity values obtained when the current electrodes are close to the potential electrodes are liable to be slightly inaccurate.

Electrode Configurations

In actual field measurements, a variety of electrode arrangements or configurations are used, the difference being in the inter- electrode distance and/or geometry. The most commonly employed configurations are the Wenner-, Schlumberger- and Dipole-Dipole arrangements.

In the Wenner electrode array, all the four electrodes, equidistant with respect to each other, are kept along a straight line, the outer two being the current electrodes (fig. 8a). The interelectrode distance is commonly denoted by the letter 'a'. The relationship for apparent resistivity, ρa, for Wenner configuration is

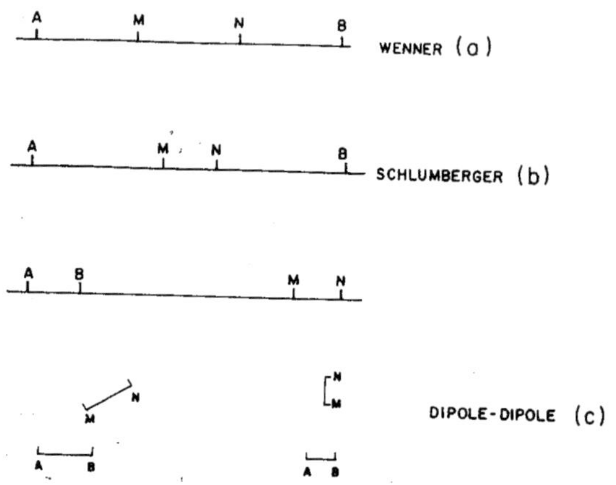

Fig. 8. Types of electrode configurations

The Schlumberger electrode configuration is also a symmetrical array like Wenner, but in this case, the potential electrodes are kept close to each other and away from the current electrodes, with the distance between the potential electrodes (MN) being generally kept less than 0.2 AB (fig. 8b). The expression for apparent resistivity for this configuration can be written as

$$\rho_a = \pi . \frac{(AB/2)^2 - (MN/2)^2}{MN} \frac{\Delta V}{I} \quad \ldots\ldots (5)$$

In the Dipole-Dipole electrode array, the current and potential electrodes are placed in pairs or dipoles (current - and potential dipoles) as shown in fig. 8c. The two dipoles may be along the same azimuth or at an angle to each other. The expression for apparent resistivity can be derived from equation (3) depending on the geometry of electrodes. The Dipole array is favoured for carrying out deep electrical sounding investigations.

Several variations of the above three arrays are also often used, such as the five-electrode variation of Wenner (Lee-partition), assymmetric schlumberger (where one or two electrodes are kept at very large distances), Carpenters array, offset array and so on. Each system has its own advantages and disadvantages.

Equipments and Accessories

Several types of equipments are available for carrying out electrical resistivity investigations. These may be grouped into two basic types : (1) those using direct current (DC) and (2) those using low-frequency alternating current (AC). While DC has the disadvantage of being affected by naturally existing currents (SP for eg), measurements with AC equipments are influenced by "skin-effect". Use of non-polarizing electrodes overcomes the problem created by natural potentials and cell-effect associated with direct current. A simple non-polarizing electrode can be designed by taking a porous earthen pot, filling it with copper sulphate ($CuSO_4$) solution in which a copper electrode is kept immersed. In certain areas such as lateritic and clay terrains, the large induced potential component retained by the formation also makes for a disadvantage for a DC instrument. Equipments available may allow one to measure the strength of the impressed current and the potential difference or the $\Delta V/I$ ratio directly. Advanced versions with attached microprocessors may take four or more readings for a single electrode separation and give the average value (Signal Averaging System - SAS). Pre- programmed equipments may even calculate the apparent resistivity values, plot the data and interpret them and print out the geoelectrical parameters, all in the field itself.

For accurate field measurements, it is preferrable to make small holes (about 5cm deep) at locations where the electrodes are to be placed or driven. These holes may then be made moist by pouring water to ensure proper electrical contact. While using non-polarizing electrodes in D.C. measurements, it may be good practice to wet the electrode positions with the same solution that is used inside the potential electrodes, particularly if S.P. measurements are also being recorded. The metal electrodes may be driven into the ground with hammers to a depth of 15 to 50 cms. depending on the type of surface formation. It is also necessary to make sure that no man- made installations (pipes, cables, etc.) exist underground in the area under investigation, since they will influence the measured potentials. The electrode spread area should be a level ground, as sharp changes in ground elevations will distort the equipotentials, hence the measured resistivity values (Fox *et al*, 1980). The electrode-spread directions must be same, as far as possible, for all measurements in an area for comparative studies. This is because of the possibility of the formations being anisotropic which will result in differing resistivity values along different electrode spread directions.

Horizontal Profiling

Horizontal profiling is done to examine lateral variations in the subsurface in the area of interest. Apparent resistivity measurements are made along chosen azimuths covering the entire area using a pre-determined electrode separation, the value of the separation dependant of the depth of interest. In practice, the whole electrode array is moved from one point of measurement to the next as a single unit. The distance between two locations where measurements are done depends on the nature and type of information sought, type of terrain, time available, etc. The resultant apparent resistivity data can then be presented in the form of either apparent resistivity profiles (fig. 9a) or Iso- resistivity contour map (fig. 9b). Generally, a good correlation can be seen between the subsurface lithology and the apparent resistivity distribution in a given area. For example, the sharp resistivity-high anomaly in fig. 9 can possibly be interpreted as a basement high

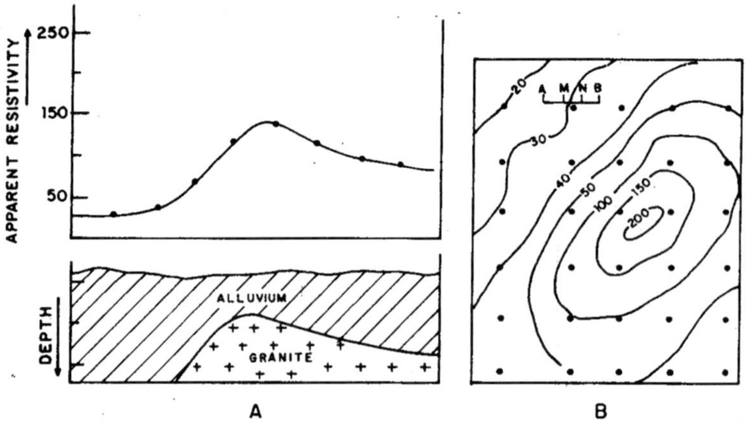

Fig. 9. Geological cross section of an area and expected apparent resistivity profile over it (A) and iso-resistivity contour map for the area (B)

Vertical Electrical Sounding (VES)

While profiling gives an indication of any changes in the lateral direction, to obtain information of the subsurface in the vertical direction, a technique known as vertical electrical sounding (VES) is employed. This involves in making a series of measurements at a given point with several electrode separations. The electrode separations are expanded, around the point of measurement, from an initial small value to several hundred meters, depending on the depth of interest. This is because, in general, larger the electrode separation, greater will be the depth of investigation. However, this relationship between the electrode separation and depth of investigation is quite a complex one, and among other things is controlled by the relative thickness of individual layers, and the resistivity contrast present between them. With Wenner array, all the four electrodes are simultaneously shifted to expanded separations from an initial small inter-electrode separation, a. For example : the progression may be, a = 1m,2, 3, 5, 7, 10, 15, 20, 25, 30, 40, 50, 70, 100m. and so on. On the other hand, if Schlumberger array were to be used, only one set of electrodes are shifted to expanded placements at a time. Normally the current electrodes are shifted to increased distances with the potential electrodes kept stationary. The need to shift potential electrodes to larger separations arises only when the signal measured becomes too small to be accurate, which may happen when the ratio AB : MN becomes very large. At such times, the potential electrodes are shifted to the next expanded position while keeping the current electrodes stationary. Generally, when the potential electrodes are shifted, at least two over-lapping measurements are made, as can be seen from the following programme.

MN = 1.0 m

AB = 5, 7, 10, 14, 20, 30 m

MN = 5.0 m

AB = 20, 30, 40, 50, 60, 80, 100 m

MN = 10 m

AB = 80, 100, 120, 140, 160, 200, 250, 300 m

The position at which the potential electrodes are shifted may be according to a previously set pattern or as and when the signal becomes too weak.

The VES data is plotted as an apparent resistivity versus electrode separation curve, normally on a log-log scale (preferrably of 62.5 mm/log-cycle). The general practice is to plot ρ_a as a function of 'a' in case of Wenner and AB/2 in the case of Schlumberger. For a site which has fairly homogeneous layers with parallel interfaces, the data obtained with wenner and Schlumberger arrays will be identical if in both cases ρ_a is plotted as a function of say, either 'a' or AB/2. The depth of investigation for different electrode configurations has been discussed by Roy (1972).

Types of VES Curves

On a homogeneous and isotropic medium the measured apparent resistivity (ρ_a) is the same as the true resistivity (ρ) of the material and therefore a VES curve obtained over such a subsurface will be a horizontal line, parallel to the abscissa (electrode separation axis) since the apparent resistivity will be same for all electrode separations. In case of a two-layered earth section, with the top layer having a resistivity of ρ_1, and bottom layer of resistivity, ρ_2 the VES curve well begin as a straight line (for very small electrode separations), rising (when $\rho_1 < \rho_2$) or falling (when $\rho_1 > \rho_2$) gradually, before again assuming a horizontal trend for large electrode separations (fig. 10). The true resistivity of the first layer is obtained by noting the initial values, for small electrode separations while that of the second layer is obtained by projecting the horizontal part of the curve on to the ordinate [app. resistivity axis], as seen in the figure. The electrode separation at which the curve rises or falls depends mainly on the thickness of the first layer - greater the thickness, larger the electrode separation value.

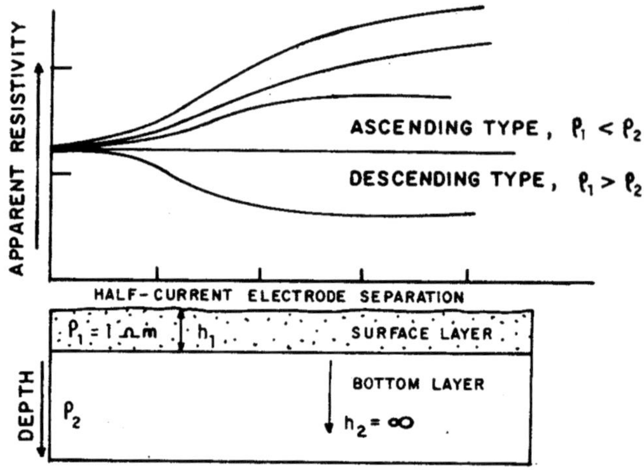

Fig. 10. Apparent resistivity curves over a two-layered earth

In areas with three subsurface layers, the VES curves may have four possible shapes. These have been named as (1) A-Type ($\rho_1 < \rho_2 < \rho_3$) ; (2) H-Type ($\rho_1 > \rho_2 < \rho_3$); K-Type ($\rho_1 < \rho_2 > \rho_3$) and (4) Q-Type ($\rho_1 < \rho_2 < \rho_3$) as shown in fig. 11. Sections having more than three layers are normally labelled by appropriate combination of the above, such as HA ($\rho 1 > \rho_2 < \rho_3 < \rho_4$), for easy classification.

Interpretation

Interpretation of the electrical resistivity data in terms of subsurface geology and hydrogeology forms the most important phase in the exploration of groundwater. The resistivity pattern and the anomalies if any, form the key in this exercise.

Interpretation of horizontal profiling data is mostly qualitative in nature. Lateral variations in apparent resistivity values are interpreted in terms of : changes in lithology (fig. 12A); variation in weathered rock/overburden thickness (fig. 12B); presence of structural features (fig. 12C); variations in the quality of groundwater (fig. 12D) etc. However, it should be remembered that, in interpreting profiling data for groundwater exploration, adequate knowledge of local geology is

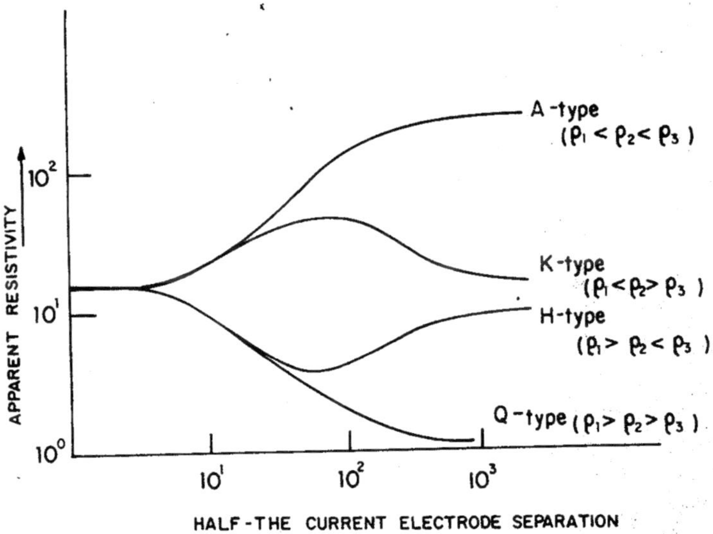

Fig. 11. Apparent resistivity curves over a three-layered earth

essential. This helps in correlating resistivity values/anomalies with specific geological units or features. Thus, in area covered by alluvium, low apparent resistivity values can be inferred as clays while higher resistivity values may be interpreted as due to sand/sandstones. On the other hand, in an area underlain by a crystalline rock such as granite, a thicker regolith/weathered rock cover will give rise to a resistivity - low anomaly. From groundwater exploration point of view, this may be the zone for siting of wells.

Profiling, normally, is a reconnaissance exercise and it is the vertical electrical sounding which gives an insight as to the actual subsurface conditions prevailing and thus help in identifying specific sites for locating water-wells. Generally, the sites for carrying out soundings are selected on the basis of profiling data - but not necessarily so.

Interpretation of VES data is both qualitative and quantitative. The type of VES curve obtained indicates the qualitative nature of subsurface that may be expected in an area. For example : a H-type Curve in a hard rock terrain may be interpreted as comprising of (1) a dry top soil cover followed by (2) moist weathered rock/regoilith, underlain by (3) the bed rock. The quantitative interpretation, on the other hand, aims at determining the resistivity and thickness of individual layers present and such other geoelectrical parameters, which are then interpreted in geohydrological terms.

Quantitative interpretation of VES data is mostly theoretical though empirical methods are still used, particularly in groundwater

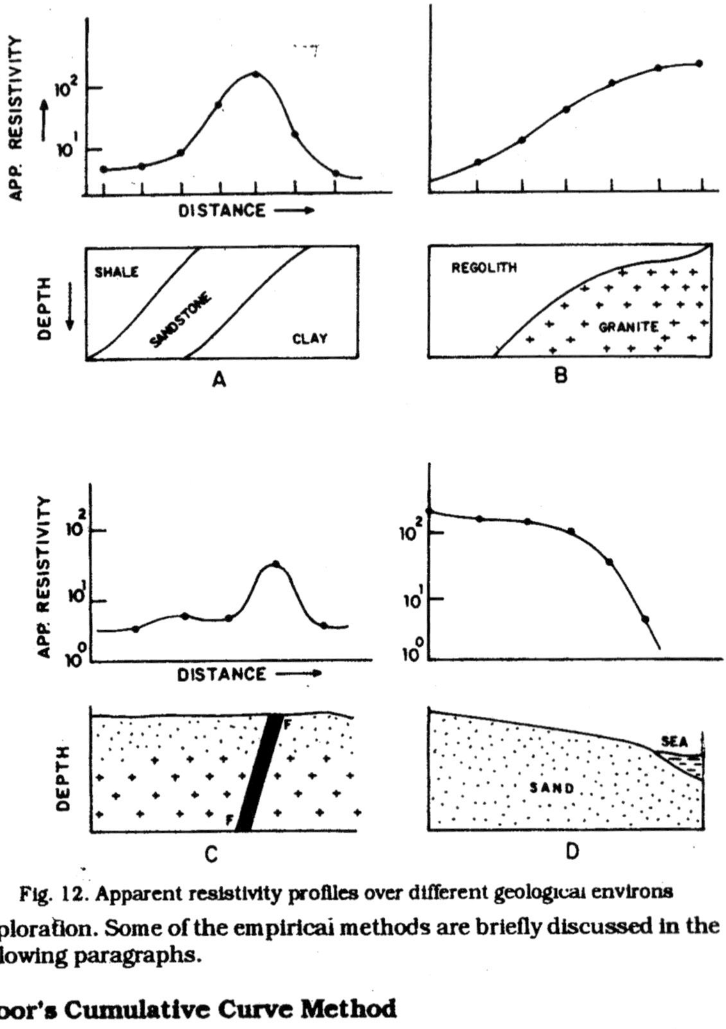

Fig. 12. Apparent resistivity profiles over different geological environs

exploration. Some of the empirical methods are briefly discussed in the following paragraphs.

Moor's Cumulative Curve Method

This is one of the oldest methods used for analysis VES data obtained with Wenner array. Here, a graph of cumulative apparent resistivities are plotted against their respective electrode separations, a, and the resulting points connected by best fitting straight lines as shown in fig. 13. The depth to layer interfaces are obtained by projecting the points of intersection of any two segments to the abscissa. Thus, it is claimed that, the number of subsurface layers present and their respective thickness are given by this simple method.

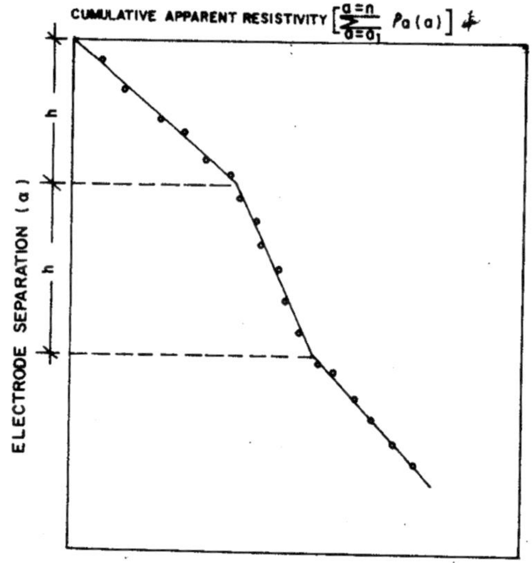

Fig. 13. Moore's cumulative apparent resistivity plot (for wenner array)

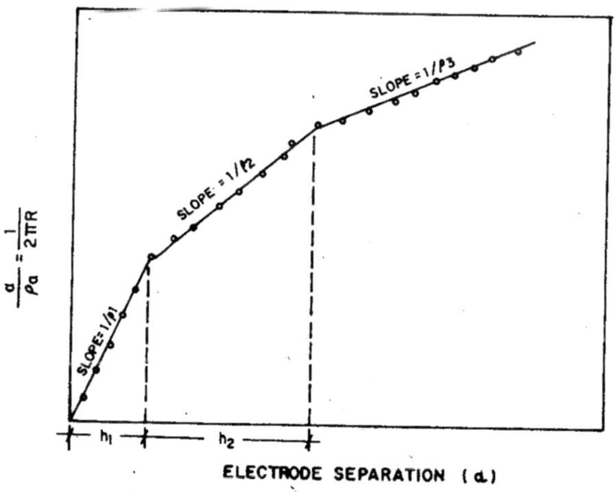

Fig. 14. Inverse-slope plot of a VES curve

Inverse Slope Method

This is a semi-empirical method suggested by Shankaranarayan and Ramanujachary (1967) for interpreting Wenner VES data. This technique is quite widely used by hydeogeologists in some parts of India, though, of late, it is being slowly discarded. Here, the values of 1/R oR ½ πR are plotted against the respective 'a' and then connected by best fitting straightline segments (fig. 14). As in Moore's method, the intersections are said to give the depths to layer interfaces while the inverse of the segment slopes give the "true" resistivities of the layer. However, this contention has been criticized by several geophysicists. An apparent draw back of the method is that when the curve slope is more than one, the method gives a negative resistivity value, which has no physical significance.

Curve-break Method

This is another empirical method suggested by Ballukraya *et al* (1983) and is also widely used in interpreting VES data for groundwater exploration. This method, again theoretically in- exact, takes into consideration the minor inflection points ("curve-breaks") in the VES curve to identify possible water bearing zones (fig. 15). Field methods have been suggested by them to differentiate between curve-breaks caused by lateral inhomogeneities and those due to horizontal discontinuities.

It is to be noted that in all the empirical methods, the depth of investigation is taken as equal to or a part of electrode separation (a, AB/2 ... etc., as the case may be). This assumption may lead to erroneous estimation of layer thickness if not applied with abundant caution.

Theoretical Methods

The theoretical methods of VES data interpretation basically involve the calculation of apparent resistivities for a series of electrode separations for a wide ranging geoelectric configurations. Thus, a geoelectric model with specified number of layers and having specified thicknesses as well as resistivities is considered and apparent resistivities for various electrode separations, with a particular electrode array, calculated. The thickness and resistivity of the first layer is always fixed at unity and those of subsequent layers are some convenient ratios. Fig. 16 gives a family of curves for a two-layered medium and for several ratios of ρ_2/ρ_1. In case of 3-layered sections, the combinations of resistivity ratio and h_2/h_1 ratios are very large. For example, a geoelectric model of a three-layer section with the resistivity ratio of 1:3:10 can be considered and a family of theoretical apparent resistivity curves calculated for different ratios of second to first layer thickness such as h_2/h_1 = 0.5, 1, 1.5, 2, 5, 10, 25 etc. Next, another geoelectric section say, having resistivity contrast of 1:3:25 may be taken up and another family of curves calculated. Thus, a large number of theoretical curves can be prepared for a wide spectrum of geoelectric sections. The field VES curve can now be matched with these theoreti-

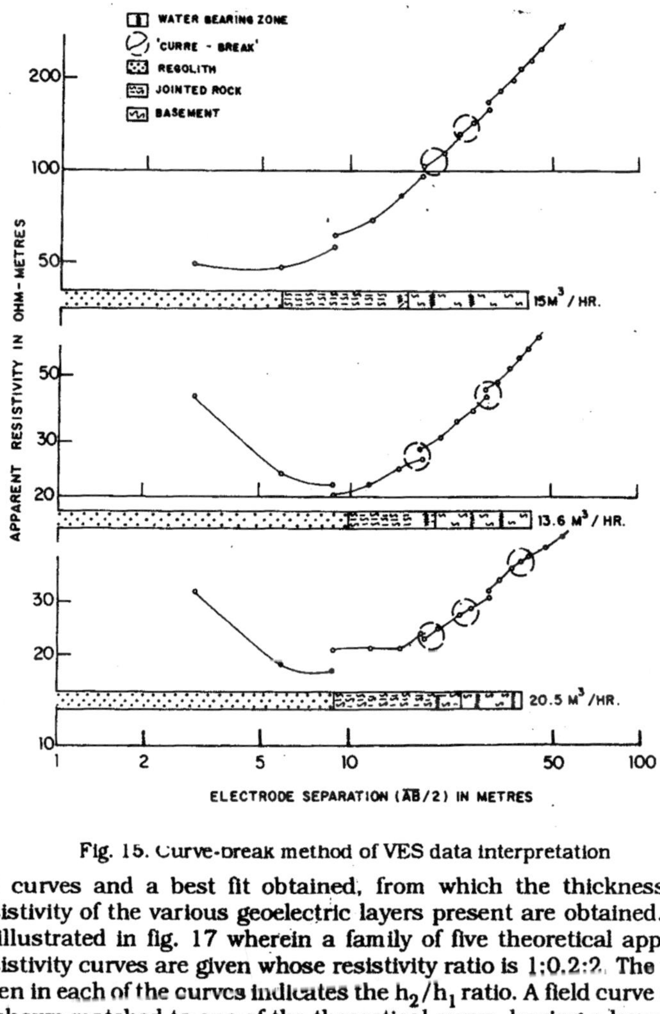

Fig. 15. Curve-break method of VES data interpretation

cal curves and a best fit obtained, from which the thickness and resistivity of the various geoelectric layers present are obtained. This is illustrated in fig. 17 wherein a family of five theoretical apparent resistivity curves are given whose resistivity ratio is 1:0.2:2. The value given in each of the curves indicates the h_2/h_1 ratio. A field curve (dots) is shown matched to one of the theoretical curve, having a layer ratio of 4. The origin of the theoretical curves is marked on the field curve as the match point. By producing this point to the apparent resistivity axis of the field curve we get the resistivity of the first layer, which is 46 ohm-meters. Similarly, the thickness of the first layer is read from the field curve's abscissa, and this is seen to be 2.7 meters. The resistivities of the second and third layers are given by the ratio 1:0.2:2, that is; ρ_2 = 0.2 of 46 Ωm = 9.2 Ωm and ρ_3 = 2 times ρ_1 = 92 Ωm. The thickness of the second layer is 4 times that of the first which is equal

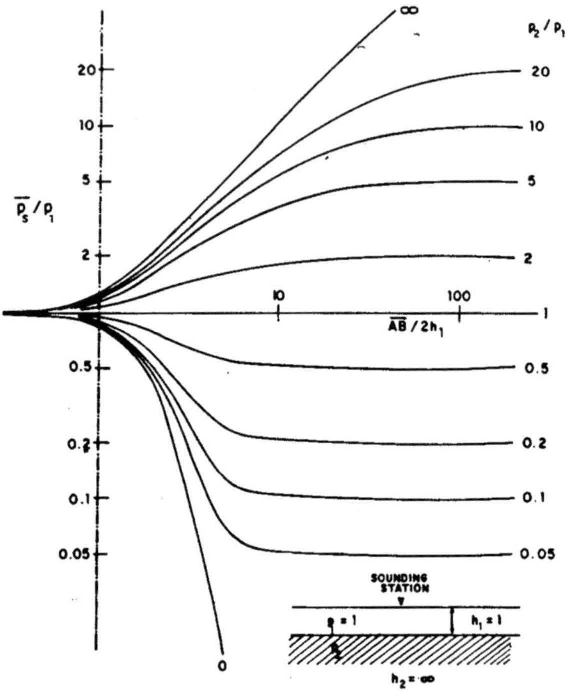

Fig. 16. Theoretical curves for a two-layered earth

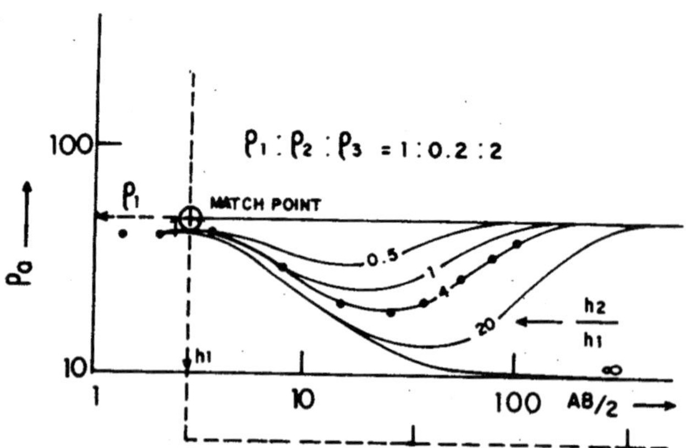

Fig. 17. Interpretation of a three-layered VES curve by curve-matching method

to 4x2.7m = 10.8 meters, the thickness of the third layer being infinity. It has to be noted that one may not always get an acceptably good match and in which case the interpretation can be done by interpolation. Several albums of these theoretically calculated apparent resistivity curves are available for ready use, viz. : Mooney and Wetzel (1956) for Wenner curves, Orellana and Mooney (1966); Com. gen. de geophysique (1955); Van Dam and Meulenkamp (1969) for schlumberger curves.

This curve matching technique, involving the search for the best match, is often a time consuming exercise. Therefore, it may be easier to carryout the exercise with the help of a computer where theoretical curves are calculated for appropriate geoelectric conditions so as to match the field curve. Several softwares are available for the interpretation of VES data with the help of desk-top computers.

Auxiliary Point Chart (APC) method

A much faster way of solving a VES Curve makes use of two-layer theoretical curves and the so-called auxiliary point charts (Bhattacharya & Patra, 1968). Two auxiliary charts are needed : H-Type for A & H-type VES curves and Q-type for Q and K-type curves. The method can be used for solving three - or more layered VES curves, since the method involves solving the curve in sections of two layers at a time.

The procedure is fairly simple, once one gets used to it. The first part of the curve (two layers) is matched with either the ascending or descending two layer theoretical curves, as the case may be and the origin of the type curve marked on the field curve sheet. The $\rho_2 : \rho_1$ ratio is also noted. This origin is superimposed on the origin of the appropriate A.P. chart, taking care to keep the axes of the two parallel to each other. The trace of the curve, for the same resistivity ratio, on the A.P. chart is now marked on the field curve as shown in fig. 18. The next part of the curve is matched with the appropriate two-layer type curve, ensuring in the process that the trace passes through the origin of the two-layer curves. Once a match is found, the origin of the type curve is again marked (second origin) and the resistivity ratio is noted. If there is one more layer, the process is continued, till the entire field curve is completed. The resistivity and thickness of the first layer is obtained by producing the first match point to the respective axes of the field curve. Next, the origin is superimposed on the appropriate A.P. chart and the value of the line passing through the second origin is noted, which gives the h_2/h_1 ratio. Thus, the second layer thickness is obtained while the resistivity of the second layer is got from the resistivity ratio noted earlier. The resistivity of the third layer is calculated by reading the value of the second origin on the field graph and multiplying this with the second resistivity ratio noted. In case there are four layers, h_3 is obtained again by keeping the second origin on the origin of the appropriate A.P. Chart and noting the value of the line passing through the third origin, which gives h_3/h_2 ratio....... In

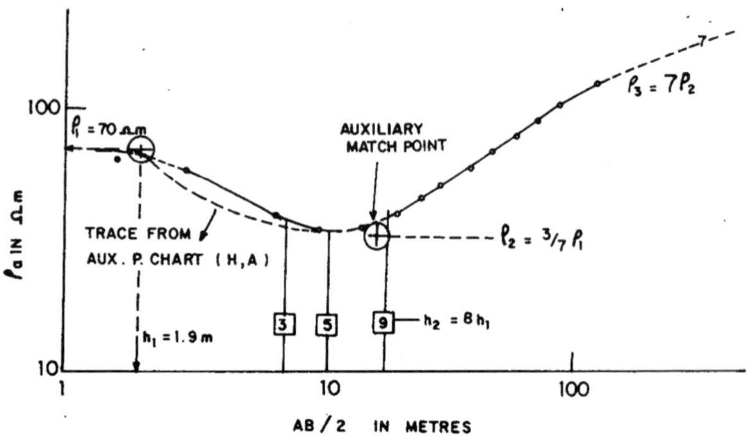

Fig. 18. Interpretation of a VES curve by auxiliary point chart method

this way any number of layers present can be solved by repeated recourse to the charts.

A quick and accurate way is to use the A.P. charts for an approximate solution and using the values so obtained, calculate the theoretical curves with the help of a computer for matching. Minor corrections can be carried out in order to get a prefect match and accurate solution.

Two other geoelectric parameters in resistivity investigations are the Total longitudinal conductance (S) and Total transverse resistance (T). The total longitudinal conductance is the conductance of all the layers of unit cross section area, the current flowing in the parallel, $S = h_1/\rho_1 + h_2/\rho_2 + \ldots\ldots h_n/\rho_n$. The total transverse resistance is the resistance offered by a series of layers (of unit cross sectional area) in series ($T = h_1\rho_1 + h_2\rho_2 + \ldots h_n\rho_n$) as shown in fig. 19.

The theoretical interpretation of VES data is predicated on two important assumptions : (a) the layer interfaces are horizontal and (b) the individual layers are homogeneous and isotropic. Apparent resistivity curves can, however, be constructed for most situations such as dipping layers, vertical and inclined contacts etc.

The solution of a VES curve is often not unique, in that two entirely different geoelectric sections can produce identical VES curves. This is due to the fact that the apparent resistivity is determined by the resistivity as well as the thickness of a layer for any given electrode configuration. Similarity of two VES curves are related to the so-called principles of equivalence and principles of suppression.

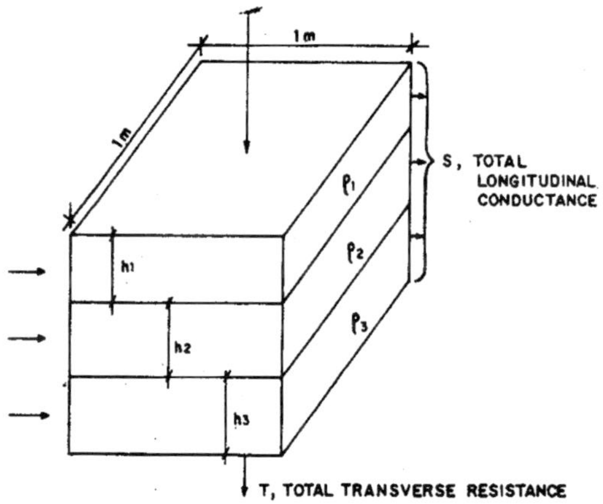

Fig. 19. Explanation of total longitudinal conductance and total transverse resistance

Two types of equivalence are generally recogneised : equivalence by T and equivalence by S.

Equivalence by T applies to K-type and to some extent Q-type curves. If two separate geoelectric sections are taken, where $\rho_1 = \rho_1'$, $\rho_3 = \rho_3'$ and $T_2 = \rho_2 h_2 = T_2' = \rho_2' h_2'$ then the VES curve in both cases will be identical. $T_2^1 = P_2^1 h_2^1$ equivalence by S occurs in H-type an also to some extent in A-type curves, if ρ_1 in section 1 is equal to ρ_1' in section 2, $\rho_3 = \rho_3$ and $S_2 = h_2/\rho_2 = S_2' = h_2'/\rho_2'$ in which case also the two VES curves will be exactly same.

Equivalence by suppression (Mailet, 1947) occurs when a thin intermediate layer in a multi-layer geoelectric section is suppressed due to small values of ratio of "relative thickness". Relative thickness is the ratio of the bed's thickness to its depth of burial below ground level. Smaller the ratio, lesser the chance of its being detected from a VES curve.

Assessment of Groundwater Availability

The presence of groundwater in an area underlain by hard crystalline rocks is dependent on the extent of weathering, fracturing and jointing undergone by the rock unit. Resistivity investigations, therefore, involve the identification of low- resistivity zones which can be correlated with weathered zones, fractured and jointed horizons etc., many of which are probably water-bearing.

In exploring for groundwater in such areas, profiling is generally done in a grid pattern, covering the area of interest, with appropriate electrode separations. Areas of resistivity-low anomalies are identified and at such locations a few vertical soundings are carried out. Interpretation of the VES data will lead to an assessment of the groundwater potential in the area.

Fig. 20A is a iso-resistivity (ρ_a) contour map, wherein a low-resistivity anomaly is seen. This may probably be due to a shear zone/fracture zone, judging from its linear disposition. Vertical soundings have been carried out in this zone and also in high resistivity zone. Two of the VES curves are shown in fig. 20B and it is seen from this that VES-1 is indicative of a thicker weathered rock and fractured rock zone as compared to VES- 2, hence better suited for drilling a well.

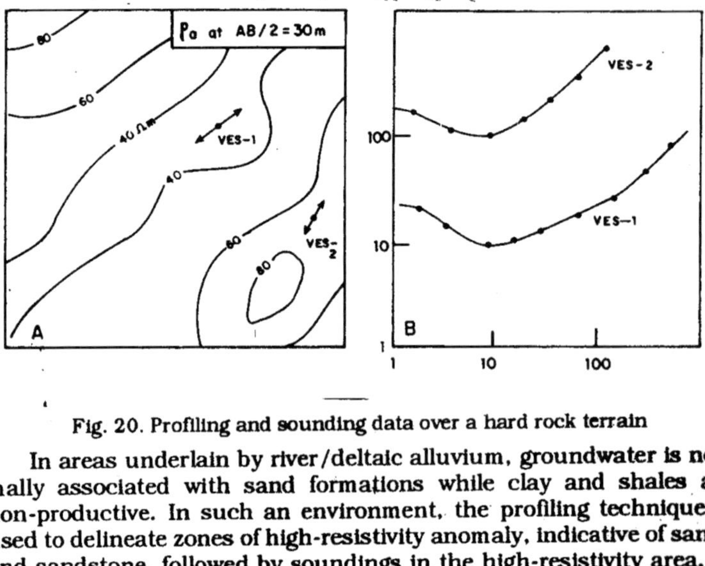

Fig. 20. Profiling and sounding data over a hard rock terrain

In areas underlain by river/deltaic alluvium, groundwater is normally associated with sand formations while clay and shales are non-productive. In such an environment, the profiling technique is used to delineate zones of high-resistivity anomaly, indicative of sands and sandstone, followed by soundings in the high-resistivity area. In fig. 21, the resistivity-high anomaly points to the existence of a sand layer/lens. This is confirmed by the VES -1 wherein a high-resistivity layer is indicated at a depth of about 20 meters or so. In contract, VES-2 does not have such a layer and in that location, the subsurface may be said to contain only clayey formations.

In areas of complex geology, where geological contacts are expected, they have to be delineated in order to properly locate a well site. Fig. 22A shows the iso-resistivity (ρ_a) map of an area where sedimentary-crystalline rock contact is expected but not exposed at the surface. The sharp increase in apparent resistivity values to the lower right corner can therefore be interpreted as due to the presence of crystalline

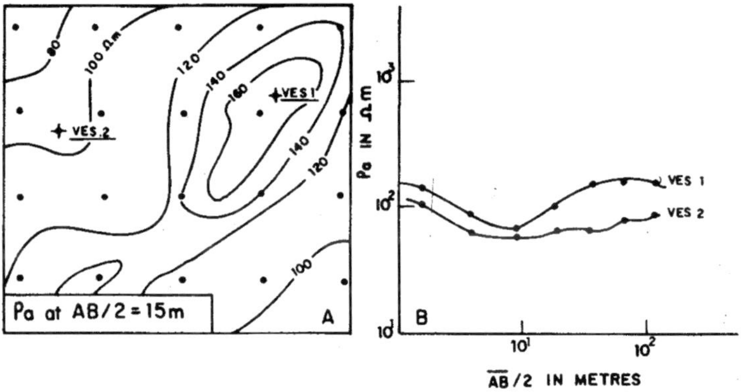

Fig. 21. Profiling and sounding data over an alluvial/sedimentary terrain

rocks (granite gneisses in this case) at shallow depths. The slope of the contact may be approximately noted from the geoelectric (ρ_a) sections (Ballukraya et al 1988) prepared from the sounding carried out in the area (fig. 22B)

In some hard rock areas, presence of lateral inhomogenities, such as a dyke, shear zone, or any other feature, have to be identified from VES curves. This is because, both horizontal discontinuities (such as

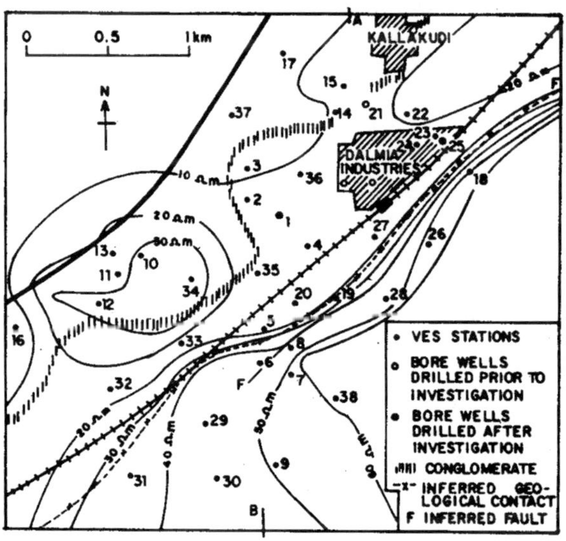

Fig. 22-A

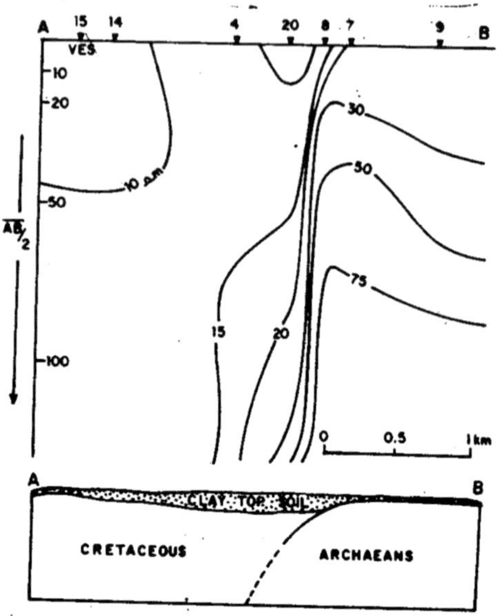

Fig. 22–B. Iso-resistivity contours (a) over a igneous-sedimentary contact and geoelectric section (b)

a weathered zone below a jointed rock layer) and lateral discontinuities (dykes, vertical/inclined shear zones, faults etc.) may give rise to identical anomalies in a VES Curve as shown in fig. 23. It is possible that such an anomaly is mistakenly taken for a water-bearing zone at depth and in such cases it is necessary to find out the exact cause - whether it is due to a horizontal discontinuity (fig. 23A) or a vertical discontinuity (B & C). This can be achieved by carrying out offset soundings and/or crossed azimuth soundings (Ballukraya *et al.* 1989).

Fig. 24 shows the result of offset soundings in an area underlain by Charnockite (Chengalpattu District., Tamil Nadu). A clearly defined low-resistivity anomaly is observed in all VES curves. The AB/2 value at which the anomaly begins in the VES curves is offset by the distance between the VES stations. If the anomaly were due to a horizontal discontinuity, then the AB/2 at which the low-resistivity anomaly occurs will approximately be the same in all the VES curves.

Profiling with Wenner array can also be used to identify lateral inhomogeneities. A sharp anomaly is obtained whenever the potential electrodes are over or near the inhomogeneity. Van Nostrand & Cook (1960) have given several examples of this in their book. However, the anomaly may be too small to be recognised if the inhomogeneity is of limited width and the potential electrodes are not near enough.

Crossed azimuth soundings, where VES are carried out along

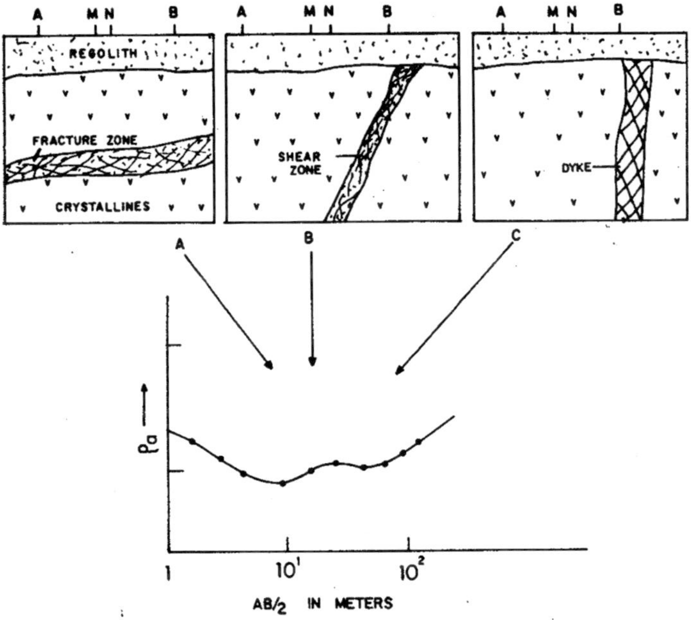

Fig. 23. Low-resistivity layer in a VES curve caused by differing geological conditions

different electrode spread directions, but about a common center are also useful in identifying lateral inhomogeneities. This data is then presented in the form of a polar diagram which gives the apparent resistivity values for various $\frac{AB}{2}$ electrode separations in different directions. Any inhomogeneity that may be present in an area is clearly brought out in these polar diagrams (fig. 25).

In carrying out electrical resistivity survey with a view to locate a well site, the following steps may be adopted to obtain best possible results.

(a) Collect all information available about the area.

(b) Map the surface geology and structure of the area.

(c) Obtain hydrogeological information of the area from existing wells.

(d) Carry out vertical electrical soundings at regular intervals (say, 50m. apart) covering the area of interest, in a grid pattern, keeping the azimuth direction common for all VES.

(e) If any low-resistivity anomalies are noticed in some of the VES curves, then, offset soundings and/or crossed azimuth soundings need to be carried out in the anomolous area.

(f) If a lateral anomaly is identified, VES can be carried out over this body, with the electrode array parallel to it, to establish the nature of the anomaly, namely resistive or conductive. (Ballukraya, 1991) as shown if figs. 24 and 26.

(g) Analyse the VES data with local hydrogeology as control and select the best possible site.

SPONTANEOUS POTENTIAL METHOD

Spontaneous potential is produced due to naturally occurring electric currents in the earths subsurface; the flow of current generated due to the presence of various minerals, water moisture, etc. Applica-

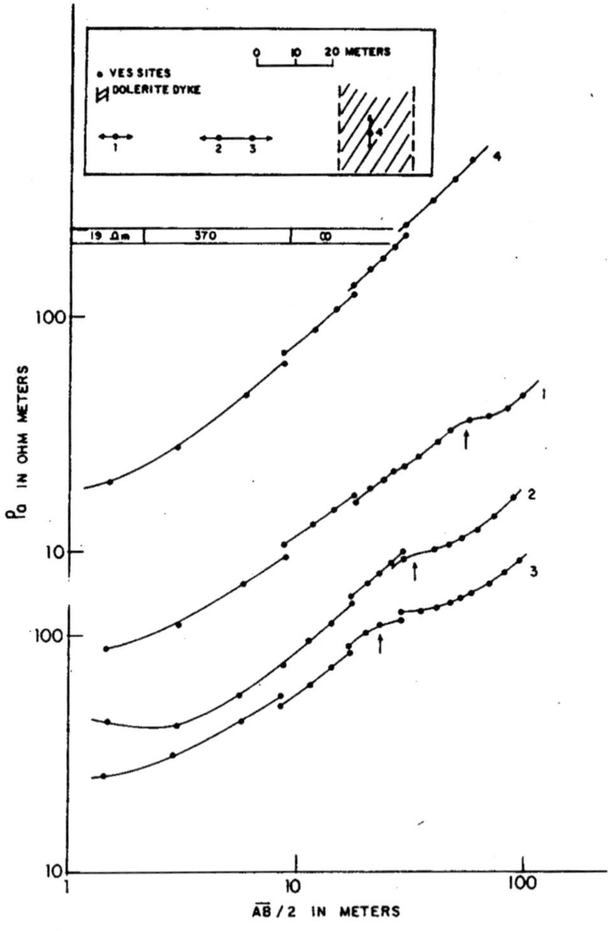

Fig. 24. Off-set sounding curves across a dyke

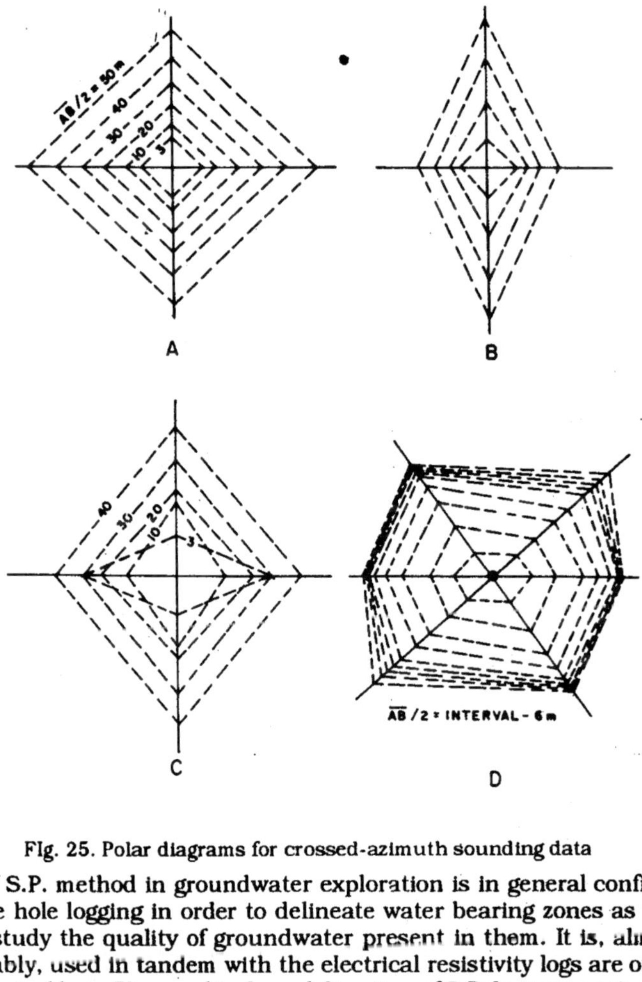

Fig. 25. Polar diagrams for crossed-azimuth sounding data

tion of S.P. method in groundwater exploration is in general confined to bore hole logging in order to delineate water bearing zones as well as to study the quality of groundwater present in them. It is, almost invariably, used in tandem with the electrical resistivity logs are other geophysical logs. The amplitude and direction of S.P. logs are controlled by the relative salinity of the formation waters and the drilling fluid. In general, if the formation water is more conductive than the drilling mud the S.P. curve has a negative deflection and vice versa.

Some attempts have been made to correlate the surface self potentials with hydrogeological parameters. One such work is by Baig (1981) who describes the relationship between the groundwater levels and the measured surface potentials. However, it appears that the relationship suggested is too simplistic since the depth to water-table

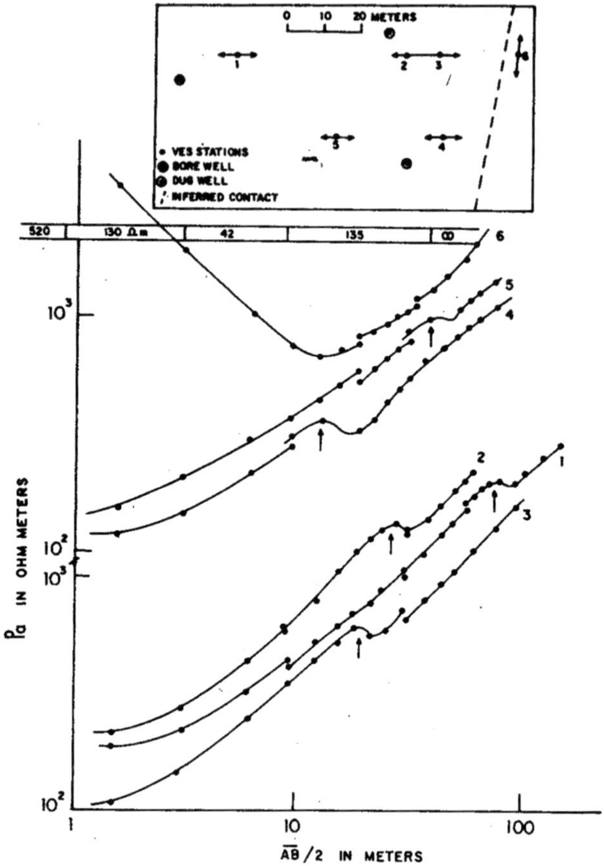

Fig. 26. Identification of conductive lateral inhomogeneity from VES data

is only one of the parameters in determining the magnitude of surface self potentials.

ELECTROMAGNETIC METHODS

This method, using artificially generated electromagnetic energy sources, is gaining popularity in hydrogeological investigations, particularly in pollution studies. It is also reported to be quite useful in locating conductive zones (fracture zones) in areas underlain by crystalline rock formations. Generally, electromagnetic waves of known frequencies are generated and transmitted with a transmitter coil, which induces electrical eddy current flow in the subsurface. The intensity of these currents is a function of ground conductivities and the secondary magnetic field caused by the eddy currents is measured

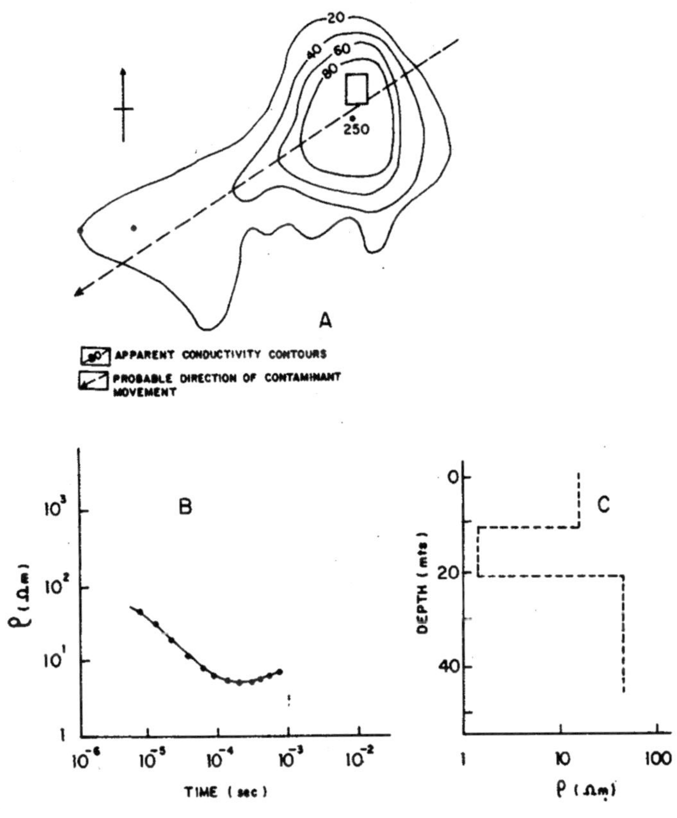

Fig. 27. Iso-conductivity map around a source of groundwater contamination (a) apparent-conductivity sounding curve (b) and inversion of this to an apparent resistivity section (c)

by a receiving coil at the surface. The value of the measured conductivity (inversely, resistivity) is influenced by the conductivity of all the subsurface layers within the effective depth of exploration. The measured value is therefore known as 'Apparent Conductivity.

E.M. methods are used either in the frequency domain (mostly for profiling) or in the time domain (for sounding). Profiling is done with a fixed transmitter - receiver distance. In the time domain EM sounding, the time-rate of decay of the EM field is measured, which is then correlated with depth. EM profiling is very useful in mapping conductive zones such as water-bearing fractures in hard rock areas; to map contaminated zones around a polluting source (fig. 27A); to identify fresh water zones in saline water areas, etc. EM sounding is similar to resistivity sounding and in fact can be converted into an resistivity Vs depth curve (fig. 27-B, C).

Another simple EM equipment that is being presently marketed by some companies consists of only a receiver coil. This makes use of radio (audio) waves as the primarily field and measures the secondary field generated in order to map conductive zones.

However, the cost of EM equipments is generally high compared to the resistivity meters. Also, interpretative techniques are much simpler in the latter case and therefore they are much favoured in groundwater exploration.

REFERENCES

Baig, M.Y.A. 1981, A simple technique of determining radius of influence in water table aquifers, *Jl. of I.W.W.A*, 13, PP. 319-322.

Ballukraya, P.N., Sakthivadivel, R. Baratan, R., 1983, Breaks in resistivity sounding curves as indicators of hard rock aquifers, *Nordic Hydrology*, 14, PP. 33-40.

--------et al, 1988, Delineation of geological contacts by electrical resistivity method, *Jl. of A.E.G.*, 9, PP. 31-36.

---------et al, 1989, Identification of lateral inhomogeneities from VES curves, *Jl. of AEG*, 10, PP. 171-184.

-----1991 *Lateral inhomogeneities - resistive or conductive ?* Abstracts of the International seminar on exploration geophysics, AEG, PP. 289-292.

Bhattacharya, P.K. and Patra, H.P. 1968, *Direct Current geoelectric sounding - Principles and interpretation*, Elsevier, 135p.

Compagnie generale de geophysique 1955, Abeques de sondage electrique, Paris.

Fox, R.C., Hohman, B.W., Killpack, J.J. and Rigo, I. 1980, Topographic effects in resistivity and induced potential surveys, *Jl of Geophysics*, 45, pp. 75-93.

Maillet, R. 1947, The fundamental equations of electrical prospecting, Geophysics, pp. 529-556.

Mooney, H.M., and Wetzel, W.W. 1956, *The potentials about a point electrode and apparent resistivity for a two-, three-, and four-, layered earth*, University of Minnesota Press.

Nettleton, L.L. 1940, *Geophysical Prospecting for Oil*, McGraw Hill Book Company.

Orellana, E and Mooney, H.M. 1966, *Master tables and curves for vertical electrical soundings over layered structures*, Interceiecia, Madrid, 150P.

Roy, A. 1972, Depth of investigation in Wenner, three-electrode and dipole-dipole D.C. resistivity methods, *geophysical prospecting*, 20, PP. 329-340.

Sankaranarayana, P.V. and Ramanujachary, K.R. 1967, Short note - An inverse slope method of determining absolute resistivity, *Geophysics*, 82, PP. 1036-1040.

Van dam J.C. and Meulenkamp J.J. 1969, *Standard graphs for resistivity prospecting*, Rijkwaterstat, The Netherlands.

Van Nostrand R.E., and Cook, K.L. 1960, Interpretation of resistivity data, USGS professional paper no. 499, Washington D.C.

9

WATER WELL–DESIGN—DRILLING—CONSTRUCTION

Dr. K. NARASIMHA MURTY

Walamtari, Himayatsagar, Hyderabad

Water Wells

In the recent past there has been increasing realisation of importance of groundwater as a source of water supply and minor irrigation. There has also been considerable development of groundwater from shallow aquifers through various types of water wells. Shallow wells are dug, bored, driven or jetted; deep wells are drilled. Increasing development in the industrial sector for manufacturing different types of shallow and medium capacity drilling rigs helped in the process of faster development of shallow aquifers. The present task is to develop deep potential aquifers so as to meet the future groundwater requirements of the country. Attention to proper drilling and design will ensure efficient wells. After a well has been drilled, it should be properly constructed, developed for optimal yield and tested.

Design

A drilled well consists of two elements, namely casing pipe and the water intake section including the gravel pack. The intake element requires hydraulic design which involves,

- mechanical analysis of bore samples collected from various depths and preparation of a bore log,
- selection of strata to be screened,
- determination of well-screen diameter, length, and durable material,
- selection of type of well, naturally or artificially packed,

GROUND WATER

Editor: P. Sankara Pitchaiah, © 1994 Scientific Publishers, Jodhpur

determination of size of gravel for shrouding.

Gradation Analysis of Aquifers

Gradation analysis is probably the most important step in the design of intake portion of a tubewell. The samples for gradation analysis are taken every 1.5-2.0 m and/or depths where the stratum changes. Considerable care has to be taken in collecting samples, particularly, when drilling by rotary method. To collect a representative sample, drill is held up and bore hole is cleared of the entire cuttings or formation material. The bit is then allowed to penetrate a known depth. All the material cut from this depth is worked out by drilling mud and collected before further drilling is allowed. The cutting is transferred to a container to decant the drilling mud. After thorough mixing, adequate quantity of representative sample is collected, dried and packed. To carry out sieve analysis about one kilogram of sample is desirable.

Grain size curve on a semi-log paper is prepared by carrying out sieve analysis of each sample collected during drilling. Commonly IS sieves from No. 75 onwards are used for this analysis (IS : 2720 - Part IV, 1965). The cumulative percent passing through each sieve is plotted on arithmetic scale and size of sieve opening on log scale. Once the plotting is complete d_{10}, d_{15}, d_{30}, d_{50}, d_{60} and d_{85} are noted. Effective diameter is given by d_{10}. The permeability (K) of the aquifer is proportional to the square of the effective grain size d_{10}, for the same uniformity coefficient. According to Allen Hazen (Taylor, 1948)

$$K = C\, d_{10}^2$$

C varies between 350-1250. The average value of C is equal to 850 if d_{10} is in mm and K in m/d and is equal to 100 if d_{10} is in mm and K in m/s. Since Allen Hazan's formula over-estimates, 2/3 of the value if K may be taken. For the same effective size, permeability is more for strata with lower value of uniformity coefficient. The uniformity coefficient is determined by the relation

$$Cu = \frac{d_{60}\ \text{(40 per cent retained)}}{d_{10}\ \text{(90per cent retained)}}$$

A uniformly graded sand has an uniformity coefficient below 4 whereas it is more than 4 for a well graded material. For gravel pack design, most investigators use d_{50} size. Sizes d15, d30 and d_{85} have also been used by some designers.

A strata chart is prepared by marking the effective size and the uniformity coefficients. This should reveal the thickness and relative permeability of each layer. In case of thick unconfined and homogenous aquifers, the screen is commonly provided in lower 1/3rd part of the aquifer. In case of confined aquifers 80-90 percent of the central depth of the aquifer is screened. Where thick heterogneous aquifers are available, it is a practice to provide screen opposite the more permeable beds leaving 0.30 cm above and below the aquifer. This is to stop the finer materials of the transition zone to enter into the well. The top of the screen should always be set below the lowest pumping level.

Design of Well Screen

The basic requirements of a well screen are presented below,

- it should have a minimum resistance to flow of water,
- it should prevent excessive movement of sand into the well,
- screen should be strong enough to withstand stresses,
- it should be resistant to deterioration and corrosion.

USBR and ISI standard 8110 - 1985 recommendations for the screen diameter for various discharges to be pumped are given in Table 1. Bennison (1947) proposed that velocity through the individual openings should be restricted to 3.0 to 7.5 cm/s in order to keep sand movement and head losses to a minimum. In India the present practice for unconsolidated formations is to allow a permissible velocity less than 3.0 cm/s.

Table 1 : Recommendations for Screen diameter

Discharge m^3/h	Screen diameter in mm	
	Minimum	Recommended
0 - 28	100	100
28 - 66	150	150
67 - 175	200	250
176 - 310	250	300
311 - 560	300	350
561 - 780	350	400
781 - 1110	400	450

Entrance velocity in State tubewells in U.P. is kept at about 3.0 cm/s. the total opening area is kept at 18 to 22 percent of total screen area to reduce the entrance velocity. The criterion for determining the diameter of well screen depends on the design yield of the tubewell. The effective open area should be taken as 50 percent of the value provided, in order to take into account the reduction in the effective area of the screen openings with time due to incrustation, rearrangement of aquifer particles around the screen and coverage by gravel, etc.

Length of the screen mainly depends upon the thickness of the aquifer available. However, minimum total length of the screen should be such that the permissible velocity through the openings is less than the permissible values given in Table 2. Procedure to calculate the screen length is illustrated in Example.

Table 2 : Optimum screen entrance velocities as influenced by coefficient of permeability

Coefficient of permeability cm/s	Optimum screen entrance velocity in cm/s
0.28	6.0
0.26	5.5
0.24	5.0
0.19	4.5
0.14	4.0
0.12	3.5
0.09	3.0
0.07	2.5
0.05	2.0
0.02	1.5
0.02	1.0

Example : The design yield of a tubewell is 125 m^3/h and a screen of 200 mm diameter having 20 percent open area is proposed to be used. Calculate the suitable length of screen, assuming the entrance velocity of water as 3 cm/s.

Solution : As $Q = A * V$

Where Q = discharge in m^3/s

A = effective open area of screen

V = entrance velocity in m/s

$$\text{Therefore, } A = \frac{125}{3600} \times \frac{1}{0.03} = 1.157 \text{ m}^2$$

Taking effective open area as 50% of that provided in order to take into account the reduction of the effective area of the screen opening with time due to incrustation, rearrangement of the aquifer particles around the screen and coverage by gravel, etc.

The required open area = 2 x 1.157 = 2.314 m^2

Open area of pipe = π x d x L x percentage of screen opening

Where d = diameter of screen

L = length of screen

$$\text{Therefore, length} = \frac{2.314}{\pi \times 0.20 \times 0.2} = 18.4 \text{ m}$$

Therefore, screen length of 18.4 m is required. Taking into account the blank portion required on both ends of screen, the screen of 20 meters length may be selected.

Materials for Screen

Water quality, cost and strength requirements often dictate the selection of material for screen. For shallow tubewells an agricultural strainer, coconut coir cord strainer, prepacked filters, and porous concrete pipes have been used effectively. For deep tubewells generally conventional slotted pipes are used. The types of well screens and slotted pipes used in India (IS-8110-1985) are plain slotted pipes, bridge slotted pipes, mesh wrapped screens, cage type wire wound screens, prepacked resin bonded gravel screens and brass screens.

In order of their ability to resist corrosion and incrustation, a list of materials is presented in Table 3. Other screen materials normally used in India are mild steel, leaded brass sheet as per IS-5494-1969, fiber glass reinforced thermo-setting plastics, PVC and copper wire, galvanized steel wire and stainless steel wire.

Table 3 : Screen materials in decreasing order of their ability to resist corrosion and incrustation.

Material	Constituents	Metal ratio
Monel metal	Nickel, copper	70:30
Supernickel	Nickel, copper	30:70
Everdur metal	Copper, silcon, manganese	96:3:1
Stainless steel	Low carbon steel, chromium,nickel	74:18:8
Silicon red brass	Copper, silicon, zinc	83:1:16
Anacouda brass	Copper, zinc	85:15
Common yellow brass	Copper. zinc	67 : 33
Amro iron	Pure iron	99.84%
Low carbon steel	-	-
Ordinary cast iron	-	-

Types of Tubewells and Slot Sizes

As discussed earlier, there are two types of wells i.e. naturally packed or developed wells and artificially packed wells. Naturally packed wells are installed in non uniform and relatively coarse aquifer (d_{90} > 0.25 mm). The finer material from the aquifer is removed during development so that only coarse material is left. In case of layered aquifer, natural development is difficult because varying slot sizes have to be made to match the corresponding aquifers.

In artificially packed wells, a large boring is needed for a casing pipe of required size. The remaining annular space is filled with graded gravel of suitable sizes. This gravel should permit use of larger slot sizes and should improve the discharge capacity of the well. It increases the effective radius of the wells and allows use of similar slot sizes in case of aquifers of different sizes. In case several aquifers of different sizes are to be tapped, slot size according to the finest aquifer should be selected provided average size of the coarsest aquifer is less than four times the finest aquifer.

The following criteria may be adopted for selection of the slot size in case of naturally packed wells (Table 4).

Table 4 : Slot size selction criteria for naturally packed wells.

Aquifer	Slot width corresponding to aquifer size
For aquifer with Cu> 6, overlying aquifers are fairly firm and will not easily cave in	d_{70}
For aquifer with Cu>6, Overlaying aquifer materials are soft and will easily cave in	d_{50}
For aquifer with Cu = 3, overlying aquifer materials fairly firm and will not easily cave in	d_{60}
For aquifer with Cu = 3, overlying materials are soft and will easily cave in	d_{40}

In case these conditions are not satisfied, each screen is to be designed with different slot sizes.

The design of a gravel pack requires determination of its average size, gradation and thickness. Though several criteria are available for design, none of them is found completely satisfactory. Investigators are now coming around the view that tubewell gravel pack should have a uniform rather than a graded grain size. Gravel pack aquifer ratio which is defined as the ratio between the average size of the gravel pack material (d_{50}) and the average size of the aquifer material (d_{50}) is generally used as a criterion for design. Based on minimum head loss through gravel pack and minimum sand movement, Central Board of Irrigation and Power, New Delhi has recommended the design criteria as

- uniform aquifer with uniform gravel pack, pack-aquifer ratio 9–12.5
- non-uniform aquifer with uniform gravel pack, pack-aquifer ratio 11–15.5.

The tubewells designed on this criteria in Uttar Pradesh have worked satisfactorily for the last several years. The recommendations on thickness of gravel pack varies from 10 cm to 22 cm. The material used for gravel pack should be clean, smooth, sound and hard with hardness number greater than 6, having less than 5 per cent of calcareous particles and should be of uniform size Cu less than or equal to 2.0. Presence of particles of shale, anhydrite and gypsum are undesirable. To avoid troubles in placing and inspection of gradation, gravel packs should not contain particle greater than 13 mm. The slot size for a gravel packed well should be such that it retains 90 percent of the gravel. Size may be reduced for corrosive waters. IS 4097- 1967 (Gravel for use as pack in tubewells) should also be consulted.

DRILLING

Methods for Constructing Shallow Wells

Shallow wells are constructed by digging, (Table 5), hand boring, driving or jetting methods and are classified according to the method of construction. Table 6 indicates the performance of the construction methods in various geological formations. Hand boring, driving and jetting methods are quite useful in construction of wells upto a depth of 45 m and to a maximum diameter of 25 cm. The drilling equipment can be conveyed easily from site to site and can be handled easily. The capital cost of equipment is negligible compared to cost of rigs. Further the drilling cost is also very much less. The space occupied by these equipment is less and they can be used where accessibility to heavier drill is not possible. Thus these methods are simple, cheaper and useful.

Table 5 : Suggested diameter and depth of dug wells

Geological Formation	Diameter of well (m)	Depth of well (m)
Hard rock areas		
granites, lime stones, Khondalites etc.	6 to 8	10–14
Soft rock areas		
Cavernous or highly weathered lime stone areas	4 (maximum)	10 (minimum)
Sand stones	3 (maximum)	10 (minimum)
Alluvium	2 (maximum)	10 (minimum)

Table 6 : Performance of well construction methods in various geological formations

Geological formation	Dug Well	Bored well	Driven well	Jetted well
Clay	R*	R	R	R
Silt	R	R	R	R
Sand	R	R	R	R
Gravel	R	R	R (fine)	R (Pea gravel)
Cemented gravel	R	NR**	NR	NR

Boulders	R	R (less than well diameter)	NR	NR
Sand stone	R (soft)	R (soft)	R (Thin layers)	NR
Lime stone	R (soft, fractured)	R (soft, fractured)	NR	NR
Disintegrated igneous rock	NR	NR	NR	NR

* R : recommended ** NR : not recommended

Drilling nethods for Deep Water Wells

Deep wells are constructed by drilling. Construction can be accomplished by percussion or direct rotary or down the hole hammer or combination or reverse rotary or dual tube methods.

Factors affecting Optimal Drilling Programme

The advances made in drilling technology and design of equipment have been more than encouraging in the recent past. Proper planning of the operations and optimum utilisation of machinery and equipment makes the task of achieving the targets relatively easy and economical. Factors required to be considered for evaluating an optimal drilling programme include :

- Complete information on the geological formations likely to be encountered during the drilling operations.
- Selection of appropriate capacity of drilling system and necessary ancillary equipment
- Moving in and rigging up operations
- Selection of appropriate sizes and types of bits required for various formations
- Selection of drill string components such as stabilizers of adequate size and number of drill collars etc.
- Determination of rotary speeds, weight on bit, mud circulation rates, hydraulics of drilling fluid etc.
- Suggested remedies for loss of circulation, abnormal formation pressures, caving problems etc. if any.
- Rig maintenance programme and maintenance schedule for all the ancillary equipment of the unit
- Operational flow chart indicating sequence of different working processes.

Character of Geological Formation

Details of the character of geological formations to be penetrated in reaching the producing horizons is essential to select proper drilling system. Basically, the geological formations are categorised into the following groups.

- Consolidated formations like hard sand stone, lime stone, trap, bassalt, quartzite, granite etc.
- Semi-consolidated and unconsolidated formations like medium to soft sand stones, shales, clay and alluvium etc.
- Bouldary formations

The drilling engineer is interested not only in those potential fresh water bearing zones but also in all other formations which must be penetrated in order to reach the producing zones. A knowledge of the composition of these rocks, their physical and mechanical properties helps in solving many of the most perplexing problems such as selection of appropriate size and type of bit, lost circulation and abnormal pressure etc. For example, consideration of cost factor alone might indicate the use of cable tool drilling system, whereas presence of abnormal formation pressures might dictate the use of direct rotary system instead, so that the drilling fluid can properly be weighted to combat the abnormal formation pressure. Similarly, where cost considerations alone might indicate the use of rotary drilling system, the presence of a zone of lost circulation might alter cost so much that the cable tool system would be preferred. Table 7 and 8 indicate the performance of the drilling methods in various geologic formations.

Table 7 : Performance of drilling methods in various types of geological formations

Type of Formation	Drilling method		
	Cable tool	Rotary	Rotary percussion
Dune sand	Difficult	Rapid	NR
Loose sand and gravel	Difficult	Rapid	NR
Quicksand	Difficult, except in thin streaks. Requires a string of drive pipe	Rapid	NR
Loose boulders in alluvial fans or glacial drift	Difficult; slow but generally can be handled by driving pipe	Difficult, frequently impossible	NR
Clay and silt	Slow	Rapid	NR
Firm shale	Rapid	Rapid	NR
Sticky shale	Slow	Rapid	NR

Brittle shale	Rapid	Rapid	NR
Sandstone, poorly cemented	Slow	Slow	NR
Sandstone, well cemented	Slow	Slow	NR
Chert nodules	Rapid	Slow	NR
Limestone	Rapid	Rapid	Very rapid
Limestone with chert nodules	Rapid	Slow	Very rapid
Limestone with small cracks or fractures	—	—	Very rapid
Limestone cavernous	Rapid	Slow to impossible	Difficult
Dolomite	Rapid	Rapid	Very rapid
Basalts, thin layers in sedimentary rocks	Rapid	Slow	Very rapid
Basalts, thick layers	Slow	Slow	Rapid
Metamorphic rocks	Slow	Slow	Rapid
Granite	Slow	Slow	Rapid

* NR : Not recommended

Table 8 : Performance of Drilling Rigs in Hard Rock Areas in South India

Drilling rig	Size cm	No. of days per well (average)	Depth of drilling per day (average) m/d
Ingersoll Rand			
TRUCM-3	15	2–3	25–32
RMT, DTH	15	2–3	9–15 (24–30)
Atlas Copco, DTH	COP 4–10	3–5	18
	COP 6–15		
Larsen Toubro, DTH	15	2–3	9–15
Vollum, DTH	15	2–3	9–15
Killick Nixon, DTH	15	2–3	9–15
Sanderson Cyclone, DTH	20	2.60	16.5
Percussion		10–15	3–4.5
Calyx	10–15	25–35	1–1.8

Selection of Drilling System

Selection of optimum size of drilling rig is governed by diameter and depth of completed well, nature of formations to be penetrated, cost of the rig, casing, bit size, speed and safety of the operation (Table

Table 9 : Selection of equipment for drilling different types of wells in various geological formations.

S No.	Type of well	Alluval formations	Mixed unconsolidated and semi-consolidated formations	Consolidated formations
1.	Boring in dugwells	Hand boring sets and light percussion rigs	Medium percussion rigs	Medium percussion rigs, down-the -hole rigs, air operated rotary-cum-percussion rock drills, calyx rigs and diamond drills.
2.	Light capacity tubewells (upto to 100 m^3/h)	Hand boring sets, light percussion rigs, light direct rotary rigs and light reverse circulation rigs.		
3.	Medimum capacity tubewells (100 to 200 m^3/h)	Medium percussion rigs, medium reverse circulation rigs and medium direct rotary rigs		
4.	Light to medium capacity tubewells		Medium percussion rigs, medium direct rotary rigs and rotary-cum-percussion rigs.	Medium percussion rigs, down-the-hole rigs, air operated rotary-cum-percussion over-burden rigs and calyx and diamond drills.
5.	Heavy capacity tubewells (200–500 m^3/h)	Heavy percussion rigs, heavy reverse circulation rigs and heavy direct rotary rigs		

9). Usually, drilling rigs are classified into light, medium and heavy duty depending on the size and depth of pilot hole that can be drilled. A broad classification followed in Central Ground Water Board, India is given at Table 10. The rig unit should also be provided with adequate ancillary equipment, transportation system and special equipment if required, so that valuable rig time is not lost on account of lack of necessary equipment.

Table 10 : Classification of Drilling Rig

S. No.	Type of drilling rig.	Classification	Dia. of hole in mm	Depth of hole in m	Size of drill rods in mm
1.	Percussion (Cable tool)	Light	150	0–50	–
		Medium	200	51–170	–
		Heavy	200	above 170	–
2.	Direct Rotary	Light	200	0–250	73
		Medium	200	251–457	73or89
		Heavy	200	above 457	89
3.	Reverse Rotary	Light	500	below 170	–
		Medium	675	above 170	–
4.	Combination (Rotary cum percussion)	*Medium*			
		Rotary	200	0–500	89
		Percussion	300	0–170	–
		Heavy			
		Rotary	200	above 500	89
		Percussion	300	above 170	–
5.	Down the Hole (DTH) Hammer	Light	114	0–50	89
		Medium	150	51–170	114
		Heavy	150	above 170	114
6.	Dual-Tube Method	—	Upto 400	Upto 400	—

Drilling string and its components

The drill string consists of swivel, kelly, drill pipe, stabilisers, drill collars and the bit. Individual drill string elements having different thread conditions are joined by means of substitutes.

The swivel performs three functions :

- it suspends the kelly and drill string
- it permits free rotation of the kelly and drill string and
- it provides a connection for the rotary hose and a passage way for the flow of drilling fluid into the top of the kelly and drill string.

The swivel must be capable of handling the weight of entire drill string. The kelly or grief stem is a hollow section whose upper end is suspended from the swivel which transmits rotation from the rotary table to the string. It may be of square, hexagonal or flute type. The tool joint thread at the lower end of the kelly is right handed and that on the top is left handed. Drill pipes compose the major portion of the drill string in deep drilling operations the top portion of the drill pipe is under considerable tension, since most of the drill string weight is supported from the derrick. The lower section of the rotary drill string is composed of drill collars. Drill collars are usually made with an essentially uniform outer and inner diameters. The purpose of the drill collar is to furnish weight and stiffness in the bottom portion of the drill string, guide and force the bit against rock formation. During drilling all the drill pipes should be in tension. Since drill pipe is essentially a tubular section and has little resistance to bending by column action, the total weight of the string of drill collars should be determined by the weight carried on the bit. However, one of the earliest recommendations was that the effective weight of the drill collars should be such that 70 percent of the drill collars would be in compression and furnishing the weight on the bit.

Selection of Bit, Drilling weights and Rotary Speeds

The type of bit which should be used is primarily governed by the characteristics of the rock formations to be drilled and the conditions under which the formation is drilled. It should also be ensured that the diameter of the bit should be enough to provide sufficient mud flow area in the annular space. Experience gained in an area facilitates selection of best suited bit for drilling operations. Economic considerations involved generally require that the hole should be drilled at the lowest possible cost. The total cost per metre depends upon, the average rate of drilling and the metreage performance of the bit. The latter becomes increasingly important at deeper depths where more rig time is spent for a round trip. The direct rotary rigs for water well drilling mostly use drag bits and rock roller bits.

Drag bits have no independently moving parts and are made of blades. The simplest of the drag bits are the fish tail bits with two blades spaced at 180 degrees apart. Bits having three or four blades are also used. The mud stream flowing out of the nozzles, positioned a short distance above the bottom of the blades, breaks up the formation, loosens it and carries upward. Drag bits are used in drilling soft formations and are quite cheaper compared to the other ones.

Rock roller bits are most widely used in direct rotary drilling. These bits have two, three or even four rolling cutters on which rows of teeth are embedded. The teeth are hard surfaced with such materials as T.C. in order to obtain longer bit life. These rollers rotate independently by virtue of their contact with the bottom of the hole. They are also provided with mud discharge ports and with reference to their size and location they are generally classified as conventional or as jet bits. Rock roller bits are designed for soft, medium and hard formations. For drilling in softer formation like soft shale, younger sandstone, etc., it is desirable to have widely spaced, long, slim teeth bits. However, the number of teeth on each roller should be slightly different from the other to prevent the teeth on successive rollers from touching the same place in the bottom of the hole. Rock roller bits for use in drilling hard formations have closely spaced and shorter teeth. This feature provides for maximum strength and tooth surface for drilling.

The primary use of a reamer is to enlarge the hole already drilled. Such enlargement is generally restricted to the practice of running a reamer above the bit while the hole is being drilled. Finally, it is to be mentioned here that the life of a bit is normally considered as terminated under one or more of the following circumstances :

- The teeth become worn out to the extent they will not effectively chip or scrap the formation.
- The gauge surfaces become so worn out that further drilling would result in an undersized hole.
- Bearing races and parts get worn out
- The rollers get jammed due to breakage of bearing races or pins.

The weight which is applied to the bit, the speed at which the bit is rotated and the weight, size and strength of drill collar are closely interrelated in rotary drilling. The weight on bit and rotary speed have got great influence on both rate of penetration and metreage performance of the bit. However, it is to be added here that minimum initial force is always required to overcome initial crushing resistance of the rock formation. Further, application of load on the bit beyond this value results in proportionate increase in the rate of rock disintegration. Field investigations in certain areas revealed faster rate of penetration and better metreage performance of the bits with large size drill collars when compared to smaller ones. This is, however, attributed to the fact that there is less buckling in large drill collars, less friction on the walls and hence virtually more weight is applied on the bit than indicated by the gauge. The effect of rotary speed on rate of penetration has been found to be difficult to evaluate. However, better rates of penetration have been observed, when the relation between weight on bit and rotary speed is maintained inversely proportional to each other. On certain occasions, it has been observed that slowing the rotary speed has resulted in an increase in rate of drilling. The following equation gives relationship :

$$HP = K \times dia \times RPM \times W \times 1.5$$

Where

$K = 0.00006$ for hard formation

= 0.00011 for soft formation

W = Weight on bit in thousand pounds

RPM = Rotary speed in revolutions per minute

HP = Horse power of the engine.

Experience has indicated that faster rates of penetration can be achieved by resorting to the combination of slower rotary speeds with greater bit weight. This practice also reduces the chances of failure of materials caused by higher vibrations due to greater rotary speeds. Table 11 gives information on selection of bit, loading on the bit and rotary speeds recommended for different rock formations. The performance of bits in hard rock areas of South India is given in Table 12.

Table 11 : Loading on Bit vis-a vis RPM of Rotary Table

Nature of formation	Structure of bit	Load in Kg per cm.	RPM of Rotary table
Soft			
Soft clay, sand, Lime stone	Long teeth widely spaced	530 to 890	100 to 60
Medium			
Lime stone, Dolomite, Hard Shale, Sand stone	T. C. teeth	710 to 980	55 to 45
	Medium length teeth with short crest moderately spaced	710 to 1430	100 to 40
Hard			
Dolomite, Chert, Sandy shale	T. C. hemispherical teeth	710 to 1070	60 to 45
Quartzite, Basalt	T. C. with strong blunt ends	1070 to 1340	60 to 45
	Short closely spaced teeth	800 to 1430	80 to 45
Hard sand stone, Trap	Short strong closely spaced teeth	1070 to 1430	70 to 50

Table 12 : Performance of Bits in Hard Rock Areas of South India

Type of Bit	Rock type	Drilling rate m/hr	Bit life hr
30 cm Ingersoll	Basalt	4.5–7.5	90–95
Rand Button bit	Granite	7.5–6.0	90–95
	Weathered granite	7.5–12.0	–
Ingersoll Rand	Basalt	3.6–6.0	90

Carset	Granite	3.0–4.5	85–90
	Weathered rock	7.5–12.0	375–400
Atlas Copco	Basalt	4.5–4.8	–
	Granite	1.0–1.5	80–110
	Limestone	5.6	–
	Laterite	3.3	–
	Quartzite	3.3	–
30 cm RMT Button	Granite	1.9	90–150
30 cm x RMT	Granite	1.5	55–130
Rollet Cone	Shale/Schist	9–12	50

Hydraulics of Drilling Fluid Circulation

All fluids used in a bore hole during drilling operations are called as drilling fluids. However, the term is usually restricted to those fluids which are circulated in the bore hole in direct rotary drilling operations. Drilling fluids in this system are pumped from the surface down through a hollow drill pipe to the bit and the bottom of the hole and returned to the surface through the annular space outside the drill pipe. Drilling fluid (mud) is a mixture of water and bentonite clay in appropriate proportions having predetermined physical and chemical properties to perform the following functions :

- Removal of drill cuttings from the bore hole
- Preventing bore hole wall from caving by providing sufficient hydrostatic differential pressure and sealing it thereby checking water intrusion.
- Holding all cuttings in suspension when circulation of drilling mud is interrupted.
- Lubricating drill pipes, bit bearings, mud pump parts etc.
- Reduce weight of drill string due to buoyancy.

Drilling, mud density, viscosity, gel strength, PH value, sand content and filtration loss properties are some of the important characteristics which govern the performance of a drilling fluid. These properties are to be frequently measured and controlled as per drilling requirements.

In normal rotary drilling operations, more horse power hours of work are utilised in circulating drilling mud than in any other single operation. The pumps, flowlines, drill pipe, bit nozzles and annular space through which mud flows deserve most careful engineering consideration especially in deep drilling operations. The rate of drilling fluid circulation is determined by the necessary upward flow velocity required for removing drill cuttings from the bottom of the hole. This upward flow velocity, also known as return velocity of mud (V) must exceed the settling velocity of the cuttings in the fluid and impart them necessary upward motion. Return mud flow velocities of 0.61 m/s to 0.915 m/s have been found to be satisfactory under ordinary operating conditions. Therefore, drilling fluid rate Q in m^3/s is given by

$$Q = A.V$$

Where A is the cross sectional area of the annular space in m^2 and V is the return velocity in m/s. The rates of drilling and circulation which are in excess of those required for maintaining the hole are used to obtain faster penetration rates.

When flushing a hole the mud pump develops a fluid pressure or a delivery head, to overcome hydraulic resistance in drill pipes, annular space, mud flow system, bit nozzles etc. The total delivery head on the mud pump, measured by means of a pressure gauge on the discharge pipe may be thought of as being composed of several components necessary to overcome particular hydraulic resistance. These components are referred to as losses of head (H) and are measured in meters of water. They are :

- Loss of head through the surface equipment such as stand pipe, hoses, swivel, kelly etc.
- Loss of head due to friction of the drilling fluid in the drill string
- Loss of head due to friction of the drilling fluid in the annular space between the drilling string and the bore hole walls.
- Loss of head through the bit nozzles

Therefore, the hydraulic horse power required of the prime mover for the mud pumps is given by

$$HP = \frac{W_f \; Q \; H}{75} \; \frac{1}{n}$$

Where W_f= The specific weight of drilling fluid kg/m^3

Q = Rated capacity of drilling fluid m^3/s

H = Total head in m

n = Efficiency of the pump (0.75 to 0.8)

Flow Chart Showing Sequence of Drilling Operations

As mentioned in the earlier text completion of a well is sum total of various working processes performed in a definite sequence. Many sets of operations can be performed simultaneously to save valuable rig time. The flow chart helps in advance planning of the well.

Economics of Deep Drilling

One of the major problems confronting drilling technology today is the cost of deep drilling operations. Though drilling costs vary from area to area but in any specific area the general relationship between depth and cost of drilling holds good. There are a number of reasons for increased per metre drilling cost as depth increases. At deeper depths, the over burden pressure on the rocks is greater, and as a result, the rocks are more compact and difficult to penetrate. Also bit life is shorter than at shallower depths. Thus more round trips are required and each of these round trips require a longer period of time. Normally, severity of the problems encountered in drilling operations become more acute at greater depths. A deep drilling rig must be designed for maximum depth at which it will be operated. Obviously, all the components are designed to meet out this requirement, which adds to the capital cost of the equipment.

Well Construction

The operations considèred as elements of well construction and completion are those items of work that are carried out after the actual drilling operations, in order to complete the well and bring it to use. These elements can be broadly classified into three categories.

- Installing the well screen in a sand or gravel formation, including artificial gravel packing, where necessary
- Developing the water-bearing formation
- Cementing and grouting the well casing in the drilled hole, sterilising the well and providing other sanitary protection.

In a few particular cases, setting the well casing in the drilled hole is also considered as a part of well construction and completion, if it is done as an operation separate from the actual drilling of the hole.

INSTALLATION OF WELL SCREENS

Alignment of Bore

Before installation of the screen it is essential to check the bore for verticality. A bore containing kinks and bends may create difficulties in installation and operation of the pump. It may cause undue wear on the pump shaft, bearings, casing and create problems in setting and removing the pump. In case of deep well turbine pumps, for casing less than 35 cm in diameter, the verticality should not deviate more than 10 cm per 30 m depth and the deviation should be in one direction and plane only (IS 2800 - 1964). If a suction air lift or a submersible pump is installed, the alignment is not so important but it is still desirable to follow the above criterion. Normally a well should be tested for verticality after drilling is completed. However, in the case of gravel packed wells, verticality should be tested after installation of the well assembly but prior to commencement of gravel filling. If the eccentricity of a bore is observed to be more than that permitted, it can be corrected by loosening earth on one side of the pipe and forcing the pipe back by applying jacks on the other side. If the hole is badly eccentric, it may have to be rebored.

Telescope-size Screens

Before discussing the various methods of setting the screens in detail, let us first consider how telescope sizes of continuous slot screens are generally installed. This type of screen is normally provided with a closed bail plug at the bottom and a lead packer at the top. Two tools generally used in setting the screen, are the screen hook and the swedge block. The hook engages the bail in the closed bail bottom of the screen either on the sand line or the hoist line, when lowering it in the well. Sometimes a string of small diameter pipe with the hook attached at the end, is employed for handling the screen. The swedge block is used to expand the lead packer to seal the top of the screen inside the well casing. In using this tool, two or three lengths of small pipe are screwed to the bar which slides through the block. This provides additional weight to the sliding bar. This sliding bar allows a driving stroke of about 30 cm. The assembly is lowered in the well and the swedge block seated inside the lead packer. The weight provided by the pipe attached to the sliding bar is then raised and dropped several times, taking care to see that the swedge block itself is not lifted out of the packer. The lead packer will expand as a result of these successive blows.

Single String Assembly

In rotary drilled wells, it is a common practice to install the casing and the screen at the same time, with the screen permanently attached to the casing instead of telescoping the screen through the casing. In such a case, the casing and screen make up a single string assembly. This assembly can be set in the drilled hole, without any bailing or washing operation, if the hole is open to the bottom. The telescope fashion of installing well screens, however, has certain advantages over the single string installation. These are :

- It facilitates the grouting or cementing of the casing in the hole before the installation of screen. Proper grouting of casing is impossible with the single string installation unless an outer casing of larger size is used.
- The telescope method also permits the screen to be removed and replaced when necessary
- In the case of a single string installation, the whole weight of the casing is brought to bear upon the top of the screen. When the screen hits bottom, it becomes a loaded column, which can easily buckle because of its slenderness, especially when the screen is long.

Several methods of installing well screens are adopted in the field, depending upon the design of the well and the method employed in drilling the well. The common methods adopted in the case of naturally developed wells are pull-back method, open-hole method, bail-down method and wash-down method. The methods generally adopted in artificially gravel packed wells are bail- down method, double-casing method and open-hole method.

Screen Installation in Naturally Packed Wells

Pull-Back Method

The well casing is first sunk to the full depth of the well, as part of the drilling operation. After the casing is sunk the hole is cleaned out to the bottom of the casing using the bailer. The telescope-size well screen is then lowered inside the casing to the bottom of the well using the sand line. If the screen is made in two or more sections, the bottom section is first lifted by a rope hitch and suspended inside the casing by means of a pair of clamps. The next section is then lifted and screwed or welded to the first one. After the screen is completely assembled, the hook is caught in the bail handle and the screen lowered to the bottom. If the depth to water is within about 10 metre's, small screens may even be dropped inside the casing. After setting the screen on bottom, the casing is pulled back far enough to expose the full length of the screen in the water- bearing formation. At the same time, care is taken to see that the lead packer is from 15 to 30 cm up inside the bottom of the casing. Using the swedge block, the lead packer is expanded to make a sand tight seal between the screen and the inside of the casing. While the casing is being pulled back, it is a good practice to rest a string of small pipe on the bottom of the screen. This provdes extra wieght to keep the screen on bottom and serves as a means of checking the exact position of the screen during the operation. The casing can be pulled back by any of the several methods commonly used. The casing may be pulled by jarring with the drilling tools. Sometimes the casing can be pulled by the casing line itself on the drilling rig. If the grip of the formations on the casing is considerable, a pair of either mechanical or hydraulic jacks will be required to provide the necessary lifting force. The jacks are set against a casing spider which grips the casing by means of wedge shaped slips.

One important rule to be observed in the case of rotary drilled wells is that when the casing is pulled back to its permanent position, it should be kept suspended at the ground surface by means of pipe clamps or spider and slips during the development of the well. Since there is little or no friction on the outside of the casing, it will not stay in place by itself until the formation is caved around it or until cement grout is placed around the casing and allowed to set. Setting the casing to the bottom of the drilled hole initially may seem to be a little extra work. However, this operation helps in preventing any difficulty from premature caving of the drilled hole which is an important factor.

Bail - Down Method

First the well casing is sunk to its permanent position, with its lower end a little below where the top of the screen will be after it is bailed down to its final position. A pilot hole may then be drilled below the casing to obtain samples of the water- bearing sand and to determine the thickness of the aquifer. The pilot hole also serves to check for the presence of boulders or other obstructions. For drilling the pilot hole, a string of smaller casing may be set inside the permanent casing. This smaller casing is pulled out once the pilot hole drilling is completed to the final depth desired. A bail-down shoe with special connection fittings is fitted to the bottom of the screen. A string of pipe called the bailing pipe is screwed on to the coupling of the bail-down shoe and the screen suspended on this string of bailing pipe. The screen is then lowered inside the casing in telescope fashion, adding lengths of bailing pipe as the screen descends, until it reaches the bottom of the initial larger hole. Using a bailer or sand pump through the bailing pipe, sand is bailed out from below the screen. As sand is removed from below the shoe by the bailer, the combined weight of the screen and the string of bailing pipe will cause the screen to settle down. Care is taken to see that the operation is stopped when the screen is bailed down to the desired depth. At this point the lead packer should be near the lower end of the casing but still inside the casing. A weighted and tapered wooden plug or lead plug is then dropped through the bailing pipe to plug the special nipple on the bail-down shoe. With this plug in place, the entire string of bailing pipe is disconnected by turning it to the right several turns to unscrew the left-hand joint at the top of the nipple. After removing the bailing pipe, the lead packer is expanded with the swedge block. In place of left-hand threaded connection at the bailing shoe for the bailing pipe, a plain-end-nipple with a socket coupling or a latch or bayonet-type connection is also used. While a screen is being bailed down, it is advisable to keep the work progressing continuously. If the work is stopped for any length of time, the formation sand may pack tightly around the screen and cause increased friction, which may prevent further downward movement of the screen altogether.

In rotary drilled wells, the well casing is first fixed in its permanent position by grouting or otherwise sealing it in the drilled hole, before bail-own operation is started.

Screen Installation in Gravel Packed Wells

Bail-Down Method

In this method, a large outer casing corresponding to the outside diameter of the gravel pack is sunk to the top of the water- bearing formation. The screen in this case is fitted with a cone- shaped bail-down shoe at the bottom. The bail-down shoe used is made sufficiently larger in diameter than the screen. The well casing and screen in a single string assembly is then lowered inside the casing. The annular space between the well casing and the outer casing is then filled with the selected gravel for some height. The screen is then jetted into position. As the screen sinks into the aquifer, the gravel being added falls around the outer surface of the well screen. After the screen reaches the desired depth, the bottom of the screen is plugged.

Double-Casing Method

This method, by far, is the most common method and best suited for most applications. The double casing method uses one string of casing corresponding to the outside diameter of the gravel pack and a second string of casing which is nominally the same size as the screen. The outer casing is first sunk to the full depth of the well. The inner casing and the well screen are then lowered to the bottom. Gravel is put in placc in the annular space around the screen. After filling a few feet with gravel, the outer casing is pulled back by a short distance. The steps of placing more gravel and raising the outer casing are repeated until the level of gravel is well above the top of the screen. The outer casing can be removed completely or the top portion of the casing can be left in place.

Placement of Gravel

The proper placement of gravel around the well screen by a good procedure is important. Simply dumping gravel into annular space from the ground surface, will result in bridging of material and in segregation or separation of portions of the material into coarse and fine layers. There are two general methods used in the field for the proper placement of gravel.

Conductor Pipe Method

When granular material is dropped in slugs within a confined space, there is less tendency for individual grains to segregate. The conductor pipe method of placing the gravel accomplished this objective. A string of 50 mm or larger pipe is installed temporarily in the annular space that is to be gravel packed. A hopper is provided at the top into which the gravel is fed. Introducing a good supply of water with the gravel will be of added advantage. Gravel may also be pumped through the conductor pipe instead of merely being fed by gravity with the water stream. The conductor pipe is raised as the level of the material builds up around the well screen. The string of pipe itself can be used to feel the top of the material placed and to measure the depth to gravel as the work porgresses. This method is practial for placing the gravel pack in wells of shallow to moderate depth.

Reverse Circulation Method

Placement of gravel by a reverse circulation system, is simpler and easier than the conductor pipe method in many cases. It can be adopted for wells of any depth, indicates the essential features of this method. As gravel is fed into annular space, circulation is set up with a descending stream in the annular space and return flow through a return pipe inside the well screen. Water is supplied to keep the borhole full as fluid is pumped out from inside the screen. The gravel is carried to bottom by the downward flow in the annular space.

Circulation is maintained by lowering the suction hose from a centrifugal pump inside the casing attached to the well screen. The water pulled through the screen is pumped back into the annular space. Some additional water may be required to be added, since there is some loss of water into the aquifer as the work proceeds. Where possible, it is better to attach the suction of the pump to a string of pipe long enough to reach almost to the bottom of the well screen. This forces the water to enter the lower part of the screen initially and insures that the first layer of gravel placed in the bottom may reach its position properly.

Recovering Screens

Occasions may arise when a well screen has to be removed from a well. Removal of the screen without at the same time pulling the casing is, or course, possible only in those cases where the original installation was made by telescoping the screen through the well casing. Occasions for pulling a well screen may be the following :

- Recovery of a screen from an abandoned well, so that the screen can be used in another well.
- Removal of a screen that has been partially destroyed by corrosion, necessitating its replacement by a new screen.
- Removal of a screen for cleaning by chemical treatment and reinstalling.

Sand Joint Method

The well screen can be readily removed from a well by a technique known as a sand joint method. This consists in sand locking a pulling pipe inside the screen solidly enough, so that when the pipe is lifted, the screen will come with it undamaged. Sand develops a high shearing strength when it is in a dense or compact condition and when it is confined so that it does not expand while the load is being applied. Both these conditions are fulfilled in the sand joint. The initial pull on the pipe gradually compacts the sand around the pipe. Then as greater pull is applied, the dense sand is prevented from expanding by being confined between the pipe and the inside of the well screen. By means of the sand joint, a lifting force of many tonnes can be applied to the pulling pipe and transmitted to the screen without deforming the screen. The procedure for making the sand joint is as follows :

- Wire some sacking just above a coupling (or a welded ring) at the lower end of the pulling pipe and cut the sacking in strips from 50 mm to 100 mm wide, depending on the size of the well (The sacking is for the purpose of forming a pocket to retain the sand fill as it is placed around the pulling pipe, the narrow strips of sacking will adjust themselves inside the screen to make the pocket when the pulling pipe is lowered inside the screen).
- Draw the top ends of the strips of sacking up around the pulling pipe and temporarily tie them in place with the pipe. Lower the pulling pipe until only the upper ends of the strips remain above the top of the well casing. Cut the string holding the upper ends of the strips, lay the strips over the top of the casing and arrange them evenly around the circumference. Lower the pulling pipe until it is near the bottom of the well screen, taking care to keep it centred in the casing.
- Pour sand slowly and evenly in the annular space between the pulling pipe and the casing. Move the pipe up and down slightly at the top while putting in the sand, in order to prevent the sand from bridging at any point in the well casing. Use a small stream of water to wash down the sand as it is being put in. Put in enough sand to fill the screen for about two-thirds of its length; but, do not overfill.
- When the sand is in place, start lifting the screen by applying tension to the pulling pipe in several steps (This reduces the chances of breaking the pulling pipe at a joint). First, apply a reasonable load and maintain it for a short time, to give the load a chance to exert a steady force on the screen. Gradually increase the pulling force from time to time until the screen begins to move, after which, continue to lift as rapidly as the equipment will permit.
- After hoisting the screen to the ground surface, loosen the sand joint by washing it with a stream of water and compressed air and remove the pulling pipe from inside the screen.

The size of pulling pipe used will depend upon the size of the screen and the pulling force required. As a general rule however, the size of the pipe used is about one-half of the nominal size of the screen. For example, in pulling a 150 mm screen, which has a clear inside diameter of about 120 mm, a line of 75 mm pipe is generally used. Table 13 indicates the sizes of pipes commonly used for pulling various sizes of screens.

Table 13 : Sizes of Pulling pipes for Sand Joins

Nminal size of Screen (mm)	Size of pulling pipe (mm)	Quantity of sand (1/m)
100	36	2.9
125	50	5.2
150	75	6.5
200	100	12.4
250	125	23.6
300	150	33.5

The type of sand which should be used to form the sand joint varies with the size of the screen and the size of the space between the pulling pipe and the inside of the screen. The most important qualities are that the sand should be clean and composed of sharp material of reasonably uniform size. The following recommendation can be taken as a general guide :

- For small size screens : medium to moderately fine sand
- For large screens : Coarse sand to fine gravel
- For very large screens : Very coarse gravel

Acid Treatment

Light acid treatment of the well will invariably facilitate easy removal of the screen. The acid loosens the rust and dissolves other incrustation, thereby helping in reducing the force required to release the screen from the grip of the formation sand. The acid treatment is done by filling the whole length of the screen with a mixture of half hydrochloric or muriatic acid and half water, using a string of black pipe or plastic pipe. The acid is allowed to stand for several hours. The well is then bailed out to remove the spent acid and any material that may have collected in the screen.

Well Development

After a well is properly constructed and gravel packed its development is taken up. Well development is the process by which the finer materials are removed from the aquifer, thereby cleaning the passages in the formation opening them up or enlarging them, so that water can enter the well more freely. The basic aim of any development operation is to produce alternate reversals of flow through the screen openings, which will result in the desired rearrangement of the particles. If the flow of water is only in one direction, small particles tend to bridge between large particles or across the screen openings. Reversing the direction of flow arrests this tendency. The process that causes these alternate reversals of flow is termed surging. When the well is surged, the outflow movement of water into the formation breaks down bridging and the inflow movement of water brings the fine materials into the well. A number of wells are to be drilled either through percussion, rotary or combination drills in alluvial/semi-consolidated and by down the hole hammer drills for drilling in hard rock areas. Development of such wells need different methods (Table 14) depending on field conditions.

Table 14 : Recommended Well Development Methods

Formation	Type of drilling	Method	Recommended method
Unconsolidated	Percussion, Combination Drill	Air lift, over Pumping, Back washing Brushing, Bailing.	Over pumping, Air Lift, Bailing
Semi-consolidated	Rotary	Surging, Back washing, Jetting, Brushing, Airlifting	Backwashing, Jetting, Surging, Airlifting
Hard Rock	DTH	Acidizing, Shooting, Chemical treatment, Hydraulic fracturing	Hydraulic fracturing is most effective and ideal

Methods

Well development methods are divided into three types, dispersal methods, concentration methods and stimulation methods. Dispersion methods include surging tool, over pumping, back washing and air lift development. Contained air lifting and high velocity jetting are grouped under concentration methods. Brushing, acidising, chemical treatment, shooting and hydraulic fracturing are included under stimulation methods.

Dispersal Methods

In this method, the development force is allowed to act on the entire length of screen at one time. Dispersion method is rather made difficult by the long screened deep wells. The methods such as surging, over-pumping, back washing tend to distribute the force over the entire area of screen. If the screen length is 30 m or more, insufficient force is available over the screen to remove much of mud cake. It has been experienced (Handa, 1983) that well efficiency over 30% is not achieved by the dispersion method of development.

With the use of surging tool, piston type movement is worked in top length of housing assembly at very slow speed. The diameter of the piston is less than the diameter of top housing. The same is done through a surge block. The movement through the formation, gives surging effect to the formation being treated. This could be done with or without chemicals. These methods become instrumental in flushing out sand, clay and finer materials which reduce the permeability of formation. Surging should only be done at very low pace. The construction of surging tool can be done at site itself by placing synthetic rubber pads between 2 steel plates, duly bolted and then joined to drill pipe connection for working with the drill for its up and down movement.

Over pumping is pumping at maximum drawdown. This method is not effective as it encourages bridging due to the flow in one direction. The operation consists of starting and stopping the pump intermittently to produce rapid changes in head. The pump is run to produce maximum drawdown and then stopped at short intervals as it has the effect of back wash through water in the pump column.

Water under pressure is forced into the formation and carries with it the clogged sand particles. The operation of back washing may be followed by bailing and then re-pumping is done till the well is fully developed and water is clear.

The compressor of adequate capacity and pressure (350 cfm and 100 p.s.i. for shallow wells and 500 to 600 cfm and 250 p.s.i. for the deeper wells) are utilised in air lift method. Compressed air is injected through air line of 1" directly through perforated foot piece in drop pipe (education line of 4") in 6"-8" or 8"-12" well assembly due to water getting lighter in weight than outside differential head. Such working of air line at different depths is continued from top to bottom for number of times till the discharge becomes sand-free and constant. Usually a compressor should run for 60/80 hours. Discharge is pulsating type and a rough estimate of discharge should be made out. Back washing should be done by stopping the discharge at random through 3 way valve intermittently and this back pushed water would move into the formation to clear it of any sand infiltration.

Concentration Methods

It is a method of choice in deep long screen water wells. The techniques most commonly and successfully employed are contained air lifting and high velocity jetting.

Contained air lifting is done between packers. On the surface, the educator is fitted with valve so that the pressure can be alternatively applied and released in chosen way. The tool isolates 1 to 3 m of the screen and concentrates the full force of air lift to the desired depth.

The operation should begin at the bottom of the screened zone to move upward slowly. Measure specific capacity of each well zone being developed to monitor the degree of development. The zone is taken as developed when specific capacity is observed stabilised.

Sometimes it is seen that well is not giving adequate discharge even when all precautions for its construction have been completed in conventional rotary drilled well. This situation might have been arisen due to the mud cake wall deposit not having been properly removed. Horizontal jetting at high velocity against the effective zones would wash the formation through screen opening increasing the well-yield and reducing drawdown. The jetting should be done with jetting tools with reciprocating type of pump fitted on the drill. This would agitate the sand and gravel particle around the screen bringing the finer ones inside the body of the well assembly itself. The same could be further alternatively cleaned with air lift as given under air lift. The jetting tools could be locally made. It is to be done with care and tact, so that at no stage screen is damaged, nor it should work over the extent desired and needed. Thus fine sand and clay would be washed off the formation giving clear discharge from the well.

Some times, in pumping water through formation, jetting and air lift development is done simultaneously. It is seen that these methods have been implemented in enhancing the discharge of 20000 1/h to 70000 1/h.

Stimulation Methods

Stimulation methods open/enlarge fractures in igneous, metamorphic and basaltic rocks and increase well yield. These include brushing, acidizing, chemical treatment, shooting and hydraulic fracturing.

Brushing involves mechanically washing with an over-size wire brush assembly and jetting the walls with high velocity water. Steel brushes are joined to the drill pipe so that it would work against the clogged portion of slotted pipe. It will result in stimulation of discharge.

Acidizing is an accepted method of water well stimulation. This increases the water production from a well. The acid-soluble-part of formation is dissloved and this permits higher rate of water to bore well. In case the pumping is being done the pumping is stopped 15% of hydrochloric acid is used in about 500 to 20000 l of water. Acid is introduced through a temporary tubing extending to a position nearer the bottom of the well and the solution is allowed to rest from 1/2 an hour to 4 days. The pump is relowered. The spent acid is pumped out of the hole in 4 to 8 hours. Acid introduced tends to flow into, widen the fractures leading into the bore wells, resulting in increased production. It is a rapid and inexpensive method.

In chemical treatment dispersing agents are added to the drilling fluid to counteract the tendency of mud to stick to sand grain. Sodium hexa metaphosphate is the best. It is added @ 1/2 1b of phosphate to 100 galons of water and allowed to stand one hour before starting development operations.

Shooting involves shooting the well with explosive detonators. This action develops local fractures around a borehole and cause agitating action which loosens the drill cutting. Shooting in rock areas and in open hole is resorted to where the discharge is meagre. This involves blasting through massive detonating charges of explosive in the form of dynamite and nitroglycerine. This has been done successfully while trying to enhance the discharge in wells drilled in limestone formations of Raipur, Bilaspur areas of M.P.

Hydraulic fracturing opens up the rock fractures so that well yield is improved. It consists of injecting liquid water into well under high pressure so that formation is actually parted or fractured. In hard rock areas where the discharge is found inadequate due to lack of water bearing cracks and fissures the hydraulic fracture can be attempted. This method is very much superior to blasting. A rubber sleeve is usually placed in the drill hole below the casing at 15 m or at desired depth from the surface and water is pumped in hole with a pressure of 50 to 55 kp/cm^2. This pressure in the production zone causes small fractures to expand. These newly opened fractures provide effective interconnection between nearby water bearing fracture and well bore. Hydraulic fracturing is not possible if there are no small cracks originally. This method has additional advantage of revitalising the sick well.

Well Completion

Contamination or surface pollution finds access to the water well mostly from the ground surface, either through the open top of the casing or around the outside of the casing. Sewage or other polluted surface waters may percolate through relatively porous upper layers of soil down to the aquifer and contaminate an area in the vicinity of the source of pollution. Due consideration and care should, therefore, be given to the selection of the well site, to possible sources of contamination in surrounding areas and to follow the accepted practices of well completion. Well completion operations can be classified into three phases.

- Grouting (or cementing) and sealing the casing
- Completing the top of the well
- Disinfecting the well

Grouting and Sealing Casing

Grouting the casing involves filling the annular space between the outside of the casing and the inside of the drilled hole with cement grout. When both an inner and outer casing are used, the grouting may be required between the two casings. The grout is a fluid mixture of cement and water of such a consistency that can be forced through the grout pipes and placed as required. Grouting and sealing the casing in water wells serves four purposes.

- It prevents downward seepage of polluted surface water along the outside of the casing
- It seals off water of unsuitable chemical quality in strata above the selected water-bearing formation
- It makes the casing stay tight in the drilled hole
- It forms a protective sheath around the casing, thereby increasing its life by protecting it against exterior corrosion.

Completing Top of Well

The top of the casing should normally extend atleast by 50 cm above the general level of the surrounding surface. In any case, the top of the casing should be well above the maximum flood water levels and should be isolated from dirt contact with accumulating drainage wastes and sudden drainage surcharges. The space around the casing should be grouted and sealed securely. A concrete platform should be constructed around the casing at the ground surface. The top of the casing should then be provided with a sanitary well seal to fill the space between the pump column pipe and the well casing. This sanitary seal consists of suitable bushing or packing gland that makes a water tight seal at the top of the casing.

Disinfection of Well

Disinfection or sterilisation of the water well including the casing, pump and pipe systems on completion of construction, is as important as any phase of construction. Practically all newly constructed wells are contaminated with direct sticking to the casing and should be disinfected promptly after completion. A chlorine solution is the simplest and most effective agent for disinfecting the well. Highly chlorinated water can be prepared by dissolving dry calcium hypochlorite, liquid sodium hypochlorite or gaseous chlorine in water. A solution containing about 100 ppm available chlorine should be used for sterilisation of the well and appurtenances. A typical procedure for disinfecting the well and its appurtenances, will be as follows:

- Determine the volume of water in the Well. (considering the well full of water will be satisfactory, since a slight overdose of chlorine does no harm).
- For each 300 m of water use 175 g of dry calcium hypochlorite with 70 percent available chlorine (or about 2.5 1 of liquid bleach with 5 percent available chlorine) which will give the desired concentration of 100 ppm chlorine solution. Mix this total amount in a small quantity of water.
- Pour this solution into the well through the top of the casing, before the well seal is installed.

- Agitate the water in the well thoroughly and let it stand for several hours, preferably overnight.
- Then flush the well to remove all of the disinfecting agent. Return the water to the well during the first part of the flushing, thereby washing the walls of the well with chlorinated water.

REFERENCES

Andhra Pradesh Cooperative Central Agricultural Development Bank Ltd. 1989 *Loan manual* (minor irrigation), 164 p.

Brantly J E 1961 *Rotary drilling hand book*. Palmer Publications, New York, 825 p.

Campbell M D and Lehr J.H. 1973 *Water Well technology*. National Water Well Association, 681 p.

Central Board of Irrigation and Power 1978 *Manual on ground Water and tubewells*. Technical report, 18, New Delhi.

Gordon R W 1958 *Water Well drilling with cable tools*. Bucyrus - Erie Company, South Milwaukee, 230 p.

Handa OP 1983 Significance of water well development in groundwater exploration in India. In : *Proc. semi. assessment, development and management of groundwater resources*, Central Ground Water Board (India), New Delhi, 1, pp. 167-177.

Handa OP 1985 *Ground Water drilling*. Oxford and IBH Publishing Co., New Delhi.

Indian Standards Institution : 2800 1964 *Code of Practice for tubewell construction*.

............: 2720 Part IV 1965 *Methods of test for soils, grain size analysis.*

............: 4097 1967 *Specifications for gravel for use as pack in tubewells.*

............: 5494 1969 *Leaded brass sheet.*

............: 7156 1974 General requirements for reverse circulation drilling rigs.

............: 7206 (1974) *General requirements for straight rotary drilling.*

............: 2800 (1979) *code of practice for construction and testing of tubewells.*

............: 8110 (1985) *Well Screens and slotted pipes.*

............: 11189 (1985) *Methods of tubewell development.*

Jain J.K. 1963 *Handbook on boring and deepening of wells*. Manager of Publications, New Delhi.

Louis Berger International, Inc. (1988) *Handbook for drainage of irrigated areas in India*. Technical report, 5, pp. 9.1 - 9.23.

Raghunath H M 1987 *Ground Water*, Wiley Eastern Ltd., 563 p.

Taylor D.W. 1948 *Fundamentals of soil mechanics*. John Wiley and Sons, New York.

United Nations Development Project 1970 Unpublished lecture notes on groundwater techniques, training course, Thanjavur, India.

Universal Oil Products 1974 Ground water and wells. St. Paul, Minn., 440 p.

Warner D. L. 1985 *Recharge wells* - 1,2. In Proc. Semi. Artificial recharge of groundwater, Ahmedabad, India

10

DEVELOPMENT OF GROUNDWATER BASINS

V. V. S. GURUNADHA RAO

National Geophysical Research Institute
Hyderabad - 500 007

INTRODUCTION

Water is not only a primary necessity of life, from time to time it also threatens life. This means water 'shows two faces' : one an aimiable one and one a malicious one. Earlier, drinking water was not a great problem because the surface waters were sufficient as well as not polluted. As urbanization and industrialization progressed the problem became more serious. Human interference in the hydrological cycle has increased considerably. Meanwhile, it became known that the various sub- systems are closely interrelated. Any intervention in one of these may affect others; for instance, pumping of groundwater often affects the surface water system.

GROUNDWATER BASIN

The term goundwater reservoir is widely used to denote places where groundwater is accumulated under conditions that make it suitable for development and use. As there has been no standard universally accepted definition of the term groundwater reservoir, it is necessary in each use to describe the particular characteristics and limiting conditions of the reservoir. "Groundwater basin", "aquifer", "water bearing zone", "groundwater district" and "groundwater area" are the terms used, applied to rocks ranging from few meters to hundreds of meters in thickness distributed over areas ranging from few hectares to thousands of square kilometers which are capable of yield to wells in the range of few liters per hour to several cubic meters per second. Conceptually the Groundwater reservoir of groundwater

GROUND WATER
Editor : P. Sankara Pitchaiah, © 1994 Scientific Publishers, Jodhpur

basin used is equivalent to the term surface water reservoir of a river basin which conveys the meaning of storing water.

A groundwater basin contains rocks that are not only porous, but sufficiently permeable to yield water by gravity to springs and wells. Generally these basins will be filled with water, sometimes in overflowing condition. The groundwater reservoir would still exist even if the entire water is pumped out, keeping the potential for storage intact and is an essential characteristic of a groundwater basin. Another important characteristic of goundwater basin is movement of the water, in the direction of decreasing head which mostly flows from recharge areas toward discharge areas. The movement of groundwater is so slow that practically all of it can be regarded as held in storage. Groundwater basins contain layers of permeable rocks surrounded by less permeable rocks. The occurrence of relatively impermeable barriers to groundwater flow makes the water in some permeable rocks becoming confined under artesian pressure. Phreatic water may flow laterally into rocks where it is confined, also confined water may leak through semi-permeable barriers and into rocks overlying them. The permeable and impermeable rocks, conduits, barriers, faults and other structural features, erosional and weathering effects, are all parts of the hydrogeologic framework of the goundwater basin. Sources and sinks of water vary largely among different basins. Generally the principal inflow is from rainfall, the outflow constitutes most of the water found in streams during dry season and groundwater pumpage, and in many groundwater basins the boundaries might be coinciding with the boundaries of the river basin, but in some cases the groundwater basins extend across the surface water divides.

For effective groundwater management, the basin should be treated as a complete hydrologic unit, including the places of natural and artificial recharge, the places of discharges and all the intervening rocks through which water may move. Conceptually, this would mean control of entire drianage basin, including surface water bodies as well as groundwater. For large river basins such a demarcation may not be practical, but it could be approached for sub-divisions of these basins (sub-basins or watersheds). For management purposes, a groundwater basin may have to be subdivided using barriers to flow and political boundaries. Such approaches are not desirable but may be required for practical reasons.

The water table represents the height to which the basin has been saturated with water in a phreatic aquifer, and changes in the water level in wells that tap the basin are direct indications of the change in the volume of water in storage. A summation of the products of the change in water level, the average specific yield of the aquifers in the zone of water level change, and the area dewatered or recharged as the case may be, gives the change in the amount of the water in storage. In confined aquifer systems, the aquifer is completely saturated and confined by overlying impervious or relatively impervious materials.

Changes in the potentiometric surface determined by piezometric measurements represent changes of pressure in the aquifer. Such changes in pressure may reflect changes in head in the recharge or discharge area, but the largest changes have been due to pumping in the vicinity of the observation wells.

NEED

What is happening is that groundwater reservoirs are coming up as quantitatively more important sources of water, though still small in total quantity compared to surface water sources as potential storage media for the surplus floodwater of wet spells to meet demands in succeeding drought periods and as potential media for controlled disposal of waste. The magnitude on which it would be advantageous to use the groundwater reservoirs in these three ways is such as would strain their potential in even larger areas. The impact of groundwater exploitation on regional aquifers is increasing and in the past, in most situations, development of groundwater on the scale needed to meet was met by reconnaissance hydrogeologic information and field visits. Now the picture has changed drastically. The reconnaissance study is still the necessary step in any comprehensive assessment of groundwater potential of any aquifer, and also the need for more detailed information have increased several fold. Now-a-days it has been difficult to give an accurate assessment of groundwater potential under complex conditions.

Groundwater reservoirs serve as large storage reservoirs, possess capacity to transmit water over long distances, with usually good chemical and bacterial quality. One of the common requirement for proper water-supply management is a knowledge of the amount or rate of change of available water in storage. A complimentary requirement is for some type of rating curve to compute how the amount in storage would change due to withdrawals from or replenishment to the storage reservoirs. The storage of surface water may be assessed with great precision on the basis of only one water level measurement for each water body in the surface water system. Because the boundaries of the surface water systems (watersheds, basins, reservoirs and lakes) are essentially dependent on topography which can be easily measured. However, where the storage and supply system relies on groundwater, the basic requirement of defining the present state of the system may be a complex problem to solve. This arises for several reasons related to the basic nature of groundwater resource, one of which is simply the fact that it is an unseen resource. Water is universally believed as a renewable resource. Renewal of water supplies is inherent in the hydrologic cycle, and it is obvious that precipitation replenishes soil moisture and stream flow. Coneptually groundwater is also recognized as a renewable resource because virtually all usable supplies are of atmospheric origin, and groundwater supplies are recharged by precipitation.

HYDROLOGIC BALANCE

The hydrologic balance is a statement of conservation of matter applied to a groundwater basin. The hydrologic balance may be expressed as follows :

Inflow		Outflow
1. Surface		1. Surface
2. Subsurface		2. Subsurface
3. Precipitation		3. Consumptive use
4. Imported water		4. Exported water
5. Decreased surface storage		5. Increased surface storage
6. Decreased soil moisture storage		6. Increased soil moisture storage
7. Decreased groundwater storage		7. Increased groundwater storage
Total inflow	=	Total outflow

The longer the time period choosen for the budget the more nearly the budget will indicate steady state conditions of the average climate. If a basin is in hydrologic balance, how it is possible to increase the amount of water taken from the area to meet human needs without destroying the long-term hydrologic balance of the basin? If groundwater is to be harnessed in an area to meet an increased local demand and the consumptive use. Such an increase can be expected to upset the hydrologic balance of the basin unless it is compensated within the basin elsewhere. In this case the only source of water to meet the increased demand comes from reduction in other consumptive uses. Annual or seasonal hydrologic budgets are evaluated and are usually made in the form of balance sheets.

A hydrologic budget defining water supply in terms of annual or seasonal precipitation is needed in order to evaluate existing and potential development of dependable water supplies in a given basin. Rainfall records are generally available for longer periods of time than the goundwater information. Available basin data on rainfall, pumping, and groundwater levels may not be available for concurrent periods. Some basic data of the hydrologic budget may be measured accurately, while others may be measured poorly or may not exist. When relatively small items in the budget have to be inferred from differences in large items, gross errors may occur because of inaccuracies in the large item data. These errors lead to poor relationships between rainfall and water supply unless the relatively small items are insignificant.

In this regard natural drainage basins are usually selected as basic units, in which to develop the hydrologic balance. Often the natural basin boundaries represent physical boundaries of the both surface and groundwater flow. Where the groundwater flow direction differs from that of surface water, it may be more useful to use the goundwater basin boundaries rather than the surface water basin boundaries. In a virgin basin, all necessary data has to be collected from available hydrologic records to evaluate the inflow and outflow terms and balance them to obtain a meaningful hydrologic balance. The process of collection of all available information is referred to as a hydrologic inventory. An inventory covering only terms required for a groundwater budget is referred to as groundwater inventory.

An independent assessment of each item of the inflow and outflow is desirable for making hydrologic budget of the basin. If some discrepancies exist in the balance, adjustment must be made within the limits of accuracy of the budget involved. If the differences in the balance exceeds reasonable limits of error in the basin data, further investigations and special designed studies may be initiated. It is rather difficult to evaluate all parameters of the inflow and outflow components of the hydrologic inventory, necessitates engineering judgment. Time lag must be recognized and adjusted in using basin hydrologic data. The hydrologic and other conditions prevailing during the period for which basic data are available must be compared with long term climatic and other conditions that are expected to prevail, and the results must be adjusted appropriately during the study period.

WITHDRAWAL OF GROUNDWATER

The amount and spatial distribution of groundwater pumping in a basin should be organized ideally so as to develop a sustainable yield on long-term. To design such a plan to accomplish safe yield of wells require the knowledge of the physical characteristics of a particular basin and their effect upon groundwater pumpage. These characteristics include the geological aspects of aquifer and aquitard layout, tranmissive property and storage capacity, their horizontal and vertical interconnections and variations, and the hydrologic aspects of source, its occurrence, movement and use of the groundwater supply.

Geologic aspects

Physical characteristics of water bearing formations :

Thickness and specific yield of water bearing deposits the greater the thickness and higher the specific yield of the deposits, the larger the storage capacity and hence larger volumes of water will be available in these deposits.

Permeability and transmissivity of water bearing deposits in and adjacent to the well fields. The higher the permeability and transmissivity of the aquiferous formations, the higher will be the capacity of these formations to transmit water to the area of heavy groundwater withdrawal in the well field.

Lithologic barriers such as relatively impervious horizontal clay layers or cemented horizons – which radically control the vertical movement of groundwater. Knowledge of clay layers is essential to understand the potential for land surface subsidence.

Structural barriers, such as faults or folds – which may impede or completely block the lateral movement of groundwater to wells. But in some situations they may increase the groundwater movement as a result of associated fracturing of the rock.

Recharge Processes

The utility of the aquifer system depends not only on the source of recharge but also on its characteristics. The sources may include rainfall, surface run-off, return flows from applied water (irrigation or waste disposal) or imported water spread from canals, etc. The supply may be from controlled releases from reservoirs and uncontrolled discharges which may occur for only short periods at erratic intervals. In basins where the magnitude of the surface water supply is extremely variable, extensive development of goundwater is advantageous and borewells should be drilled to draw maximum groundwater from storage. The amount of water that recharges a groundwater basin is a limiting factor which decides the amount of water that can be developed. This quantity depends to some extent upon the area available for recharge. Natural recharge in shallow water table zones is limited by impervious surface developments. However for confined aquifers, the groundwater supply is limited by the rate at which water can be transmitted from recharge areas and by the rate at which water can pass through confining layers in these aquifers.

WATER USE

Water use has to be distributed by appropriate placement of wells closer to the demand areas. Under natural conditions water table gradient and goundwater movement will tend to conform with the surface topography which in many inland basins, results in drainage at the downdip side in a basin. Legal and water quality aspects must be considered in detail in any basin development. The legal problems may result from the fact that reduction of water levels in the outflow regions will reduce the quantity of water naturally draining from the basin, effecting the downstream supply. The quality problems are related to the fact that a decrease in outflow will tend to create adverse salt balance conditions, particularly if pumped water are applied within the basin. In coastal regions the aquifers outcrop on the ocean floor. Under natural conditions some subsurface outflow occurs into the ocean. In some basins. the aquifers extend to considerable distances into offshore region and store large volume of fresh water. Development of these basins, alters water levels and if the levels are declined below sea level saltwater may begin to enter the aquifer system which decreases the water quality.

GROUNDWATER QUALITY

Maintenance of groundwater quality at acceptable levels is a major requirement for fruitful utilization of aquifer system. The low velocity of groundwater flow, which inhibits mixing and diffusion within a groundwater basin is one of the principal factors effecting groundwater quality. Leaching of aquifers is usually expensive and also impervious materials form localized barriers to groundwater flow and obstruct uniform dispersion of water throughout the aquifer. Consequently some variations in water quality under natural conditions can be expected to occur in various parts, both horizontally and vertically within the aquifer system. Most significant characteristic of groundwater quality deterioration is the fact that it is usually a cumulative process. It is easier to prevent than to purify, once pollutant concentrations have reached problem proportions. Often, maintenance of water quality becomes a constraint in groundwater basin operation and experience proved that many aquifer systems suffer quality deterioration over a period of time. Thus one should plan in advance on how to deal with degradation from various sources.

MONITORING PERFORMANCE OF TOTAL WATER RESOURCE SYSTEM

Integration of groundwater management plans into the total water system and, as far as possible into the total environment, should be considered during formulation of the plans. The first step toward integration is to rationally combine the groundwater and surface water facilities to permit their co-ordinated use. The co-ordination of groundwater basin involves the use of a basin for transmitting and storing varying portions of the area's water supply by co-ordinating the aquifer functions with surface facilities such as reservoirs and pipe lines, to meet the water requirements of an area. Significant attributes of groundwater basins are the amount of fresh groundwater in storage and the basin's storage capacity and transmission characteristics. Alternative plans of conjunctive operation should be studied to assure the optimum use of these attributes to satisfy the physical needs for water.

Water level elevations

Water level data can best be presented in the form of groundwater contour maps including contour maps of changes in water levels. From these maps, the direction of groundwater flow, the influence of pumping, the influence of recharge, and special effects caused by underflow, effluent seepage, sea water intrusion, etc. can be identified. Once the conditions and characteristics of the aquifers are known, the number of data points used in preparing the maps can be minimized. Hydrographs of the water levels in a well are also useful in dipicting variations in level at a particular point over the period of record. A general inventory of groundwater conditions may be determined from contour maps drawn once a year. These maps are generally prepared from data obtained at the time of annual high groundwater levels or of annual lowest groundwater levels following the pumping season. The main objective of water monitoring program is periodic evaluation of the water resource management system.

Water quality

In formulating an effective plan for control and maintenance of groundwater quality, full consideration must be given to geology, hydrology, and cultural development within the area of the groundwater reservoir. Water quality characteristics must be established and trends observed. A comprehensive investigation program for evaluation of these factors is essential. Groundwater should be sampled regularly on a continuing basis. Sampling points should include shallow and deep wells, selected with due consideration to areal distribution, geologic and hydrologic influences, pattern of pumpage and waste disposal practices. Samples should be obtained when water levels are at their lowest and also, if possible, during the period of replenishment when water levels are high. In coastal aquifers particular emphasis should be given to changes in chloride concentration in groundwater. Total dissolved solids content is important where salt balance problems exist. In general, a groundwater monitoring program should be used to point out emerging problem areas, so that timely corrective action can be initiated. The collection of basic data should form an integral part of a comprehensive program involved in protection and maintenance of groundwater quality. Data over an extended period are necessary to determine the sources and sinks, behavior of contamination and protection of a groundwater basin.

Waste disposal considerations

Often groundwater quality problems stem from the fact that people live and carryout their activities on ground surface overlying groundwater basins. Inevitably many of the wastes that result from these activities will find their way to the groundwater underlying the ground surface. Because of the great harm due to the various contaminants entering groundwater and difficulties involved for correcting a polluted groundwater condition, there is a growing interest in the topic of groundwater contamination and its control. Evaluation of the effects of a waste discharge on groundwater quality is unique and site specific. Factors include the local hydrological conditions, type of contaminant, its relations with bacteria and with the soil media and the hydraulic conditions governing the velocity of the groundwater, the plume shape and dispersion.

Data Base

The measurement of the upper surface of the saturated zone

requires access to the water table through dug wells or borewells. Additionally, the groundwater levels change significantly from place to place, thus many water level measurements may be required to define the water table configuration and changes in the amount of groundwater in storage. Because of the fact, the rate of change in groundwater systems are generally small, thereby regional trends can often be assessed through either annual or seasonal measurements. Additional difficulties crop up from the fact that the boundaries of groundwater basin do not necessarily coincide with topographic highs or drainage divides. Thus the definition of the lateral boundaries of a groundwater basin may require extensive test drilling. Even if defined at one time, groundwater divides may be transitory in nature, as the divides are prone to migration in response to temporal changes in recharge to and withdrawal from a groundwater system.

Further, the assessment of groundwater resources are complicated by the three dimensional nature of groundwater flow. The bottom of the aquifer is commonly undefined because aquifer properties tend to be more poorly known at greater depths. Adding to the difficulty of defining the groundwater resource is the significant spatial variability in the storage and transmissive properties of rocks. In general, porosity and permeability tend to decrease with depth. The hydraulic conductivity varies from point to point because of the inherent spatial variability of the geological factors that control it (such as grain size, shape, sorting, packing, orientation and fracturing). The hydraulic gradient is the change in hydraulic head per unit distance in the direction of flow. The state variable of hydraulic head, which governs hydraulic gradient and thereby groundwater flow rate, can be measured in the field through wells, borewells or piezometers.

THE SYSTEMS APPROACH

The increasing complexity of groundwater development and groundwater information gathering is part and parcel of the vastly larger picture of increasing impingement of the activities of man on each other and on their environment as their numbers increase, their desires multiply, and the resource on the earth are exploited to satisfy them. We have to recognize the effects of interactions and to accept the necessity of studying the entire environment as a dynamic system subject to analysis. Later it can be utilized for planned management. The best way to assess the hydrologic system or part of it that is of interest, assess how it functions, decide what knowledge is necessary to understand and explain its functioning, update the knowledge that would be needed as a basis for altering its functioning in desired ways, and finally look at what knowledge is available to see what is missing. Herein specifically to groundwater, we must look the groundwater segment of the overall hydrologic cycle and also at its boundaries or interfaces with other segments. We must try to analyse all the hydraulic, chemical, and biological processes that are involved, their interaction with others and with the boundaries, and the changes in the processes and interactions that may occur in response to actions of man or events of nature. This systems approach brings things we do not understand and the kinds of studies we must make to understand them well enough to manage them appropriately.

Remote sensing is being tested for usefulness in many types of studies–hydrologic, geologic, geomorphic, agriculture, etc. It can cover large areas, and provide information no other technique can produce with almost synoptic coverage of the whole earth. Groundwater resources offer advantages of different type in many developing countries. In hot climates evaporation from the surface of a reservoir. can be extremeldy high, accounting almost one-third of the flow into the reservoir. If a groundwater resource is to be exploited, it is essential that the entire project is carried out in the most efficient and cost-effective way possible. The wells must be designed so as to be efficient, and that the pumps must operate at near peak efficiency, but also implies that every step of the operation must be carefully planned and implemented in a predetermined logical order. If a projected groundwater programme is carefully thought out in advance, a reliable supply of water should be obtained assuming that this is possible hydrogeologically. The evaluation, development and management of an aquifer for water supply should proceed in stages. The simplest and less expensive phase of the investigation should be undertaken first. The data obtained will have to be used to decide the way forward to the next phase.

ESTABLISHMENT OF DEMAND

The foremost in an investigation should be to establish the quantity and quality of water required from the groundwater development. The present as well as projected demands and the ultimate use of the water are important to determine the quality of water to be supplied in a planned way. To meet the future requirements high potential 'source has to be investigated firstly and it has to be further investigated for pollution hazards in future'

FEASIBILITY STUDY

Desk study in volves gathering of all the relevant data concerning the study area. The main objective is to estimate :

– the location, extent, and thickness of potential aquifers, distance between the source and demand,

– depth to watertable,

– transmissivity and storage properties of aquiferous formations,

– approximate estimates of perennial yield based on meteorologic and hydrologic data.

Groundwater exploration

A groundwater investigation requires a thorough appreciation of the hydrology and geology of the concerned areas and groundwater inventory needs to determine possible gains and losses affecting the groundwater reservoir. The information concerning lithology, stratigraphical sequence and geological structure as well as the hydrogeological characteristics of the subsurface formations is essential. The location of watertable heights and their fluctuations are equally important. The records of rainfall, climatological conditions usually provide supplementary information in groundwater investigations. Stream flow records also help assess the gain or loss to the aquifer system.

Groundwater management

The quantity of water available from an aquifer depends mainly on the amount of natural recharge, the rate at which groundwater is withdrawn by wells and the underground storage available. The three stages in the development of an aquifer are :

(i) Initial development stage when groundwater pumping is much less than natural recharge only spring and river-flows are affected marginally.

(ii) Full development stage when groundwater abstraction equals natural recharge over long years. The effect of groundwater pumping on the surface water bodies may be critical during later stages of development if the aquifer is not managed efficiently.

(iii) Over development stage when groundwater pumping exceeds natural recharge.

REFERENCES

American Society of Civil Engineers, 1987 *Groundwater Management*, ASCE Manuals and Reports on Engineering Practice No. 40.

Hamill, L and Bell, F.G. 1986 *Groundwater Resource Development*, Butterworths. pp. 343.

11

ASSESSMENT OF GROUNDWATER BASINS

V. V. S. GURUNADHA RAO and C. P. GUPTA

National Geophysical Research Institute,
Hyderabad - 500 007

INTRODUCTION

Lack of fresh water to drink, for use in industry and agriculture and multitude of other purposes the most important factor hindering development in many parts of the world. As populations increase, the demand for water grows. While the world's freshwater resources are large, they are not only unevenly distributed in both space and time but also have a finite limit in terms of practical utilization. So the shortages currently affecting many areas are likely to spread, becoming for more prevalent during the next decade. In many regions these shortages are likely to be exacerbated by climate change.

Under these circumstances, improving knowledge of the world's water resources, particularly groundwater resource, is indispensable for the well being of mankind and for the protection of the environment. Reliable information on the state and trends of groundwater resources is a prerequisite for sound decisions of their sustainable management. Information is required for water resource planning, design of projects, impact assessment, project operation and for forecasting or warning of extreme events such as droughts. Hence groundwater resources assessment is a precondition for all other aspects of water resource development and management.

Groundwater resources assessment is a continuing long-term activity, as it must be to support it's management that is sustainable into the future. Particularly the environmental impacts of human activities, there is an increasing need for reliable information. This implies present and future generations will benefit from decisions made

GROUND WATER
Editor : P. Sankara Pitchaiah, © 1994 Scientific Publishers, Jodhpur

today and the information generated by today's water resource assessment activities. If sustainable development is to mean anything such development must be based on an appropriate understanding of the environment–an environment where knowledge of water resources is basic to virtually all of man's endeavours.

Groundwater resource assessment is the determination of the sources, extent, dependability, and quality of water resources on which is based our evaluation of the possibilities for their utilization and control. The expanding human activities are becoming increasingly intensive and diverse and are having an ever-growing and more evident impact on natural resources through depletion and pollution. Changing world climate effects the distribution of rainfall, run-off and groundwater recharge. It cannot be assumed that the future patterns of these hydrological phenomena will continue to be as they were in the past. Only with reliable data and information on the status and trends of the groundwater resource-including quantity, quality, statistics on extreme events and use can help arrive decisions on how best to manage it.

DATA NEED

Most groundwater models are structured to solve a partial differential equation or a set of partial differential equations for groundwater systems. The data requirements for aquifer modeling is presented in Table 1. The following elements need their complete description (Javandel et al, 1984).

Table 1. Data requirements for groundwater modeling (after Moore, 1979)

I. Physical Framework

A. Groundwater Flow

1. Hydrogeologic map showing areal extent, boundaries, and boundary conditions of all aquifers
2. Topographic map with drainage pattern and surface-water bodies
3. Water-table, bedrock configuration, and saturated thickness maps
4. Transmissivity map showing different aquifers and boundaries
5. Map showing variation in Storage coefficient
6. Relation of saturated thickness to transmissivity
7. Hydraulic connection of stream/river with the aquifer

B. Solute Transport (in addition to above)

8. Estimates of the parameters that comprise hydrodynamic dispersion

9. Effective porosity distribution
10. Background information on water quality (natural concentration) in aquifer
11. Estimates of fluid density variations and relationship of density to concentration
12. Hydraulic head distributions (to determine groundwater velocities)
13. Boundary conditions for concentrations

II Stresses on System

A. Groundwater Flow

1. Type and extent of recharge areas (irrigated areas, recharge basins, recharge wells, etc.)
2. Surface water diversions
3. Groundwater pumpage and its spatio-temporal distribution
4. Stream flow (distributed in time and space)
5. Rainfall

B. Solute Transport (in addition to above)

6. Areal and temporal distribution of water quality in aquifer
7. Stream flow quality (distribution in time and space)
8. Sources and strengths of pollution

III Other Factors

Groundwater Flow and Transport

1. Economic information of water supply
2. Environmental factors
3. Planned changes in water and land use
4. Legal and administrative rules

- The distribution of the parameters through the space of interest.
- The initial conditions for variables,
- The sources and sinks of interest and their variation with time, and
- The boundary conditions at the margins of the space of interest.

Flow parameters

For groundwater flow problems the parameters of aquifer of interest are porosity and permeability. Compressibility of both the fluid and the medium are mathematically combined as specific storage or in the case of unconfined aquifer specific yield.

Sources and Sinks

Recharge or discharge through wells, other natural/artificial recharge or discharge.

Boundary conditions

Impermeable boundary, or prescribed flux, prescribed value of the state variable, viz., head or pressure. Often a combination of boundary conditions will apply.

Geology

An understanding of geology of aquifer systems helps in defining the geometry and distribution of the parameters in space as well as boundary conditions. Modeling of real systems is carried out by simplifying and abstracting information on these systems as parameter distributions and boundary conditions, which helps in tackling the problem mathematically. It is desirable to select the appropriate mathematical model which preserves this conceptual model.

GROUNDWATER FLOW MODELS

A groundwater flow model is defined as a non-unique, simplified version of a real groundwater system, portraying the features essential to the purpose of which the model was developed and incorporating various assumptions and constraints pertaining to the system. A model expresses the conceptual representation of the system and between the system and its environment. It describes these relationships both in terms of mathematics and uncertainity, facilitates simulation of system behavior under various conditions. The simulation envisages prediction of system responses with assumed system parameters and known stresses, evaluation of system parameters when history of stresses and responses, for the system parameters are either known or bounded (Konikow and Mercer, 1988).

The solution of the mathematical model can be either continuous or discrete in space and/or time. Numerical models are based on numerical approximation of both space and time. According to the usuage, the models are categorized as prediction models, which simulate the behavior of the groundwater system and its response to stresses. Resource management models, which integrate hydrologic prediction with explicit management decision procedures and Identification models, which estimate input parameters for the above two.

Groundwater flow models utilize information on aquifer parameters, boundary and initial conditions and human interventions to solve the mathematical equations for studying the quantitative aspects of groundwater flow such as flow direction and flow rate, changes in stream-aquifer interactions, etc. Land subsidence models portray the phenomenon of land subsidence which is caused by heavy groundwater abstractions of various pumping schemes. Mass transport models address the question of groundwater quality by incorporating the mathematical approximation of the transport viz., fluid flow (convection) and/or mixing of one or more chemical constituents in the groundwater. These predict the movement and concentration in the aquifer of various pollutants including the radionuclides, leached soils from land fills and irrigated areas, and saltwater intrusion in coastal aquifer.

The philosophy underlying a deterministic modeling approach is that given a high degree of understanding of the processes by which stresses on a system produce subsequent responses in that system, the system's response to any set of governing processes even if the magnitude of the new stress falls outside the range of historically observed stresses (Konikow and Pattern, 1985). The accuracy of the deterministic prediction depends partly, on how best the concepts of the governing processes are understood which are controlling the system behavior. However, even if an accurate conceptual model of the governing process exists, and in turn, the processes are represented accurately in a simulation model, the following are vital for :

(1) definition of the properties and boundaries of the domain over which these stresses are acting,

(2) the state of the system at some point in time and

(3) an estimate of what the future stresses will be.

The deterministic models are developed using the governing processes. These are generally described as a mathematical equation or a set of equations. The prediction comes from the solution of the governing equations and phenomenological laws . The number and types of equations are determined by the concepts of the involved governing processes. The coefficients of the equations are the parameters that are measures of the properties, boundaries and stresses of the system, the dependent variables of the equations are the measures of the governing equations. The spatio-temporal variation of groundwater system is obtained in terms of the governing mathematical equations, that are normally written in terms of the change in the dependent variables with respect to both space and time. The prognosis of the state of the system at some future time can be viewed as the production of the computed rates of change in the dependent variables from some defined or known state at a point in time represents the initial conditions which must always be specified for solving a transient condition.

Groundwater hydrologists are aware that data inadequacy and insufficiency limit the reliability of deterministic simulation models. The data may be inadequate because

(a) aquifer heterogeneities occur on a scale smaller than can be defined from the available data

(b) time-dependent variables are monitored too infrequently and

(c) due to existence of inherent measurement errors.

Consequently, there is lot of research work carried out for better representation of errors or inadequacies in data in a stochastic manner. Further these statistical concepts are incorporated in an otherwise conceptually-based deterministic modeling approach. Freeze (1982) and Newman (1982) present overviews of stochastic groundwater modeling and also contrast stochastic and deterministic approaches.

For groundwater flow problems it is reasonable to assume that isothermal conditions prevail and that fluid properties (density and viscosity) are homogeneous. With these simplified conditions a generalized form of the groundwater flow equation describing the non-homogenous anisotropic aquifer may be derived by combining Darcy's law with the continuity equation (Bear, 1979). The dependent variable in groundwater flow equation is hydraulic head. However, in cases where fluid properties such as density or viscosity vary significantly in time and space because of changes in pressure, temperature or chemical composition, the fluid is non-homogenous and relations among water levels, pressures and fluid velocities become complex. Computation of flow rates and directions then require data on pressure, density and elevation in addition to head measurement (Konikow and Patten, 1985). In most of the groundwater studies, it can be reasonably assumed that groundwater flow is two dimensional. Thus a three dimensional flow equation could be reduced to a case of two- dimensional areal or cross-sectional flow, for which several additional simplifications are applicable. The reduction of dimensionality of the equation helps use of less stringent data and simpler efficient mathematical means to obtain a numerical solution.

For obtaining a unique solution of the governing partial differential equation describing the given physical process the information of the physical state of the processes is needed. This data is provided by initial and boundary conditions. For steady state problems, only boundary conditions are required, whereas for transient problems, both boundary and initial conditions are needed. Mathematically the boundary conditions include the geometry of the boundary and the values of dependent variable or its derivative normal to the boundary. For groundwater flow problems the boundary conditions are generally of three types viz., specified head, specified flux or head dependent flux (Mercer and Faust, 1981). The initial conditions could be simply the values of the dependent variable specified everywhere inside the boundary.

NUMERICAL SOLUTIONS

Once the formulation of the mathematical model is complete, the next step is to obtain a solution using one of two general approaches viz., analytical or numerical. Generally, the problems of groundwater flow are complex and analytical models no longer can describe the physics of the situation, partial differential equations can be approximated numerically. In doing so, the continuous variables are replaced by discrete variables that are defined at grid blocks (or nodes/meshes). Thus the continuous partial differential equations describing the groundwater flow are replaced by a finite number of algebraic equations that define hydraulic head at specified points. The resulting system of algebraic equations is generally solved using matrix techniques. This approach constitutes a numerical model and generally, a computer code is written to solve the equations on a computer.

Two versatile approaches of numerical techniques for groundwater flow equation are finite difference methods and finite element methods. Each of these numerical methods consists of a variety to sub-classes and implementation alternatives (Mercer and Faust, 1981). Both the numerical approaches require that the area of interest be sub-divided into a grid or a number of smaller sub-areas that are associated with node points. Finite difference methods approximate first derivatives in the partial differential equations. In both approaches, the descritization of space and time allows the continuous boundary value problem for the solution of the partial differential equation into a set of simultaneous algebraic equations, suitable for solving by using either iterative or matrix methods. Each approach has advantages and disadvantages but there are very few groundwater flow problems for which either is clearly superior.

MATRIX SOLUTION TECHNIQUES

As elaborated above, each numerical approximation leads to an algebraic equation for each node. These are combined to form a matrix equation, which may be solved numerically by direct or iterative methods. Direct methods require storage of large matrices and introduce lot of round off errors, and convergence and the rate of convergence are of concern. Numerous iterative schemes have been developed, a few of more commonly used ones include successive over relaxation methods, alternating direction implicit procedure, iterative alternating direction implicit procedure, strongly implicit procedure (Remson et al, 1971) and point successive over relaxation method (De Marsily, 1986). Generally the iterative methods start the solution with an initial estimate, the efficiency of the method depends on the initial guess. To hasten the iterative process, relaxation and acceleration factors are used. The iterative approaches demand error tolerance to be specified to stop the iterative processes. This too may be problem dependent. Recently, semi-iterative method is gaining popularity (Gresho, 1986) and is known as 'conjugate gradient method.' The advantage of the conjugate gradient method is that it does not require the use or specification of iteration parameters, there by eliminating this partly subjective procedure (Manteuffel et al, 1983).

UNCERTAINITIES OF PARAMETER ESTIMATES

The major cause of errors in groundwater models output is the presence of errors or uncertainties in the input data set which reflect the aquifer properties, stresses, and boundaries. To facilitate model calibration and to quantify model reliability, the parameter estimation techniques have been used in conjunction with groundwater flow models. In parameter estimation the hydrologic experience and judgment of the modeler continues to be a major factor during calibration procedure of a model. The computation procedure of parameters for groundwater flow consist of finding a compatible set of parameters (transmissivity, storage coefficient, and input/output stresses) that minimize deviations between observed and computed values of hydraulic head. Most of the parameter estimation techniques require initial estimates of aquifer and their statistical properties. This has stimulated an interest in obtaining the statistical properties of parameters directly from field data. One of the more promising procedures is krigging which is a stochastic interpolation technique. Delhome (1979) provides a good account of the methodology and case studies in which groundwater models were used in conjunction with krigging.

MODEL CALIBRATION

To help demonstrate that a deterministic groundwater simulation model is realistic, field observations of aquifer responses are compared to corresponding values obtained from the model. The objective of this procedure is to minimize differences between the observed data and computed values, it is commonly known as calibration. Usually, the model is considered calibrated when it reproduces historical data within some acceptable level of accuracy. A poor match indicates evidence of errors in the concepts (or hypotheses) underlying the simulation on the other hand a good match in itself does not prove the validity or adequacy of the model. The calibration of a deterministic groundwater model is often accomplished through a trial-and-error method of model's input data to modify the model's output, which is a subjective procedure. The hydrological experience and judgement of the modeller continues to be a major factor in calibration. The modeller should visit the field area being aware of the local conditions. He must also recognize that the uncertainity in the specification of sources, sinks, and boundary and initial conditions should be evaluated during the calibration procedure in the same manner as the uncertainity in aquifer properties. One of the objectives of model calibration is to improve the conceptual model of the aquifer. The conceptual model consists of our understanding of the physical and functional nature of the aquifer and the simulation model numerically integrates the effects of the many factors the affect groundwater flow or solute transport, the computed results will be internally consistent with all input data, and one can determine if any element of the conceptual model must be revised.

In most field studies there are some inadequate data, so the values for some parameters must be estimated. A common approach is to first assume the best estimates of values for parameters based on available measurements and then adjust the values until a best-fit is achieved between observed and computed dependent variables. Another objective of the calibration procedure is to determine the sensitivity of the model to factors that affect the predicted magnitude of the variables of interest, such as groundwater head or solute concentration. Evaluating the importance of each factor helps determine which data must be defined most accurately and which data are already adequate or require only minimal additional definition. If additional data cannot be collected, then the sensitivity tests help to assess the reliability of the model by demonstrating the effect of a given range of uncertainity or error in the input data on the output of the model. The relative sensitivities of the parameters that affect flow and solute transport vary from problem to problem and sensitivity analysis is a must and vital during the early stages of a model study. (Konikow and Patten, 1985).

MODEL VALIDATION

Increasing concern regarding the validation of the computer models has come up during recent years. The question that occurs, even after a computer code is verified, that is, the code is mathematically correct in its formulation and solution, is whether the computer model is valid in making the proper predictions. This is an important issue to those who need to rely on modeling results to make long-term management decisions. The term model validation may be defined as follows :

"A conceptual model and the computer code derived from it are validated when it is confirmed that the conceptual model and the computer code provide a good representation of the actual processes occurring in the real system. Validation is thus carried out by comparison of calculations with experimental measurements and field observations."

A few important factors that are used during model validation to simulate physical processes in geologic formations are : (1) The model is no more than a conceptual picture that the modeler constructs for the system, together with the corresponding mathematical equations and their numerical solutions. The results depend strongly on how it's inputs are designed and how it is used. Since the conceptual picture can never be as detailed as the real system because of the lack of information and computer capacity, the model is inadvertently a simplification of the reality. If the model representation of the real system is adequate, it may yield results for specific observables of interest with the required accuracy and within a specified range of conditions. Thus, it becomes necessary to embark on the following before modeling exercise starts :

(1) the observable quantities of interest

(2) the accuracy required for the prediction of these quantities and

(3) the range of conditions for which the model is validated.

It is important to establish the range of applicability of validated models. which cannot be developed that are valid for all situations. Carefully defining ranges of applicability can perhaps help to avoid applications of the models to conditions for which they have not been validated. In modeling processes of geologic media, it is useful to differentiate between processes and model structures. Model structures represent geologic and geometric characteristics of the medium, such as faults, layering, and heterogeneity. Processes are physical and chemical phenomena such as groundwater flow, colloidal transport, and dissolution and precipitation. Processes can be represented by mathematical equations, whereas model structures are site-and scale-dependent and are part of the input data to the model. For a reliable modeling study, one needs both the proper process identification and the proper model structure selection. Failure of modeling results to match field data can be due to errors in either. Processes and model structures are approached quite differently. Ideally, models of processes can be validated generically, but models of geometric structures are site-specific. However, in practice one may notice an intimate relationship between processes and model structures.

Therefore, one should say that a model, including the conceptualization and the code, is validated with respect to (a) a process or (b) a site-specific system. Initially, a process is first identified and then conceptualized and later coding is done. The model, composed of the conceptualization and code, is then applied to a carefully designed study for which the controlling process and its results are compared with observations. A site-specific system may be composed of a number of processes and structures which form the building blocks of the system. It is an important step, which may involve substantial efforts, to identify them. Once the building blocks are identified, a model or a group of models may be used to simulate a site and be compared with filed observations. If successful, the group of models is said to be validated with respect to this particular site, within a range of applications determined through observations. Hence, there is no such thing as a vlidated model in the generic sense. However, one can say that 'a model is validated with respect to given physical process,' or that 'a model or group of models is validated with respect to a given site.'

CASE STUDY (After Narasimha Reddy et al., 1992)

Parkal watershed, in crystalline rocks of granitic terrain, covering about 12.35 sq.km. is situated in Warangal district, Andhra Pradesh State and falls under Semi-arid Tropics. The watershed is bounded by Lat. 18'10'45" to 18°13'05" N and Long. 79°39'30" to 79°42'20" E

(Fig. 1). The rainfall mostly occurs during South- West monsoon from June to September and the mean annual rainfall is 1090 mm. The groundwater divide coincides with the topographic boundary, thus forming a closed groundwater regime. Some outflow occurs across 2.61 km section of the Paidpally Tank in the North. Both excess surface water as well as groundwater leaves the watershed across this section. The streams are emphemeral with intermittent flash flows after good rains and the surface runoff is being stored in three tanks.

Sandy Loams and Loamy sandy clays occur in the watershed with thickness ranging from 0.3 to 1.0 m. Paddy is the only irrigated wet crop grown in the ayacut lands of tanks. Maize and green grams (cereals) are the major rainfed crops grown during monsoon season. The second crop is mostly grown in the ayacut of tanks and is also supplemented with groundwater resources which in turn may contribute irrigation return flow to the groundwater regime.

HYDROGEOLOGY

The watershed is underlain by crystalline rocks of Archean age comprising of grey and pink granites and traversed by quartz and pegmatite. Weathering of rocks has been observed down to a depth of 8 meters below ground surface. Fractures at depths, below the weathered zone have been identified during drilling. Groundwater occurs under water table conditions in the weathered and fractured parts of the hard rocks.

Groundwater levels in the open wells and borewells generally start rising from the last week of June till the first week of October. There are twenty observation wells monitored regularly since 1980. The depth to water level during October varies between 0.5 to 7.0 m.bgl whereas during premonsoon the water levels stand at 2.5 to 9.0 m bgl. Recharge to the groundwater regime mainly takes place from rainfall.

A. RECHARGE PROCESS MODEL

The water balance method of estimation of groundwater recharge utilizes the balance among various components of the water balance equation.

$$RE = P + Ir - (Int + Rof + AE + ATR + WC)$$

Where P = Rainfall

Ir = Irrigation

Int = Interception loss

Rof= Surface runoff

AE = Actual evaporation from soil surface

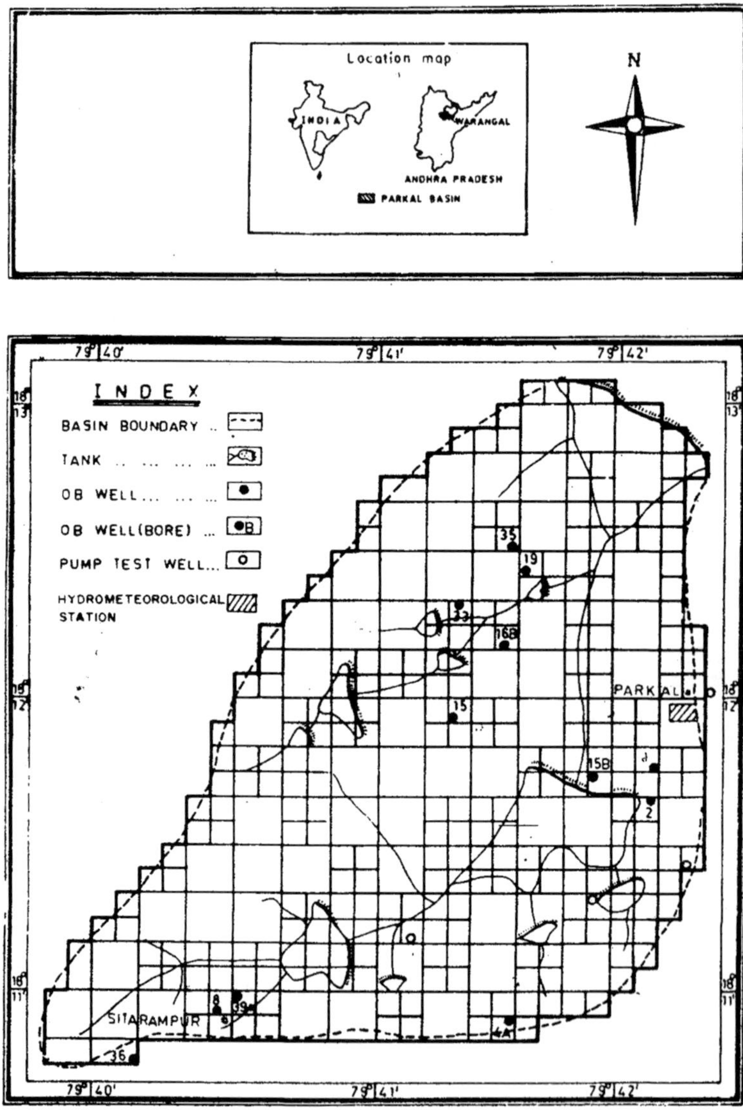

Fig. 1. Grid map of meshes with stream channels, watershed boundry and location of observation and pump tested wells.

ATR= Actual transpiration

WC = Soil moisture storage and

RE = Groundwater recharge

The daily rainfall data has been used to estimate the surface runoff whereas pan evaporation measurements have been utilized to estimate actual soil evaporation and transpiration from the vegetation.

Interception Loss : The interception loss in the study area has been assumed as 0.5 mm per rainfall event on a rainy day preceded by a dry day. If the precipitation for this rainy day is less than or equal to 0.5 mm, the interception loss is equal to the precipitation. Also for a rainy day fallowed by previous rainy day, the interception loss has been assumed to be zero.

Surface Runoff The surface runoff from the rainfall has been estimated through use of soil conservation service (SCS) runoff Curve Number Model (Ven Te Chow, et. al. 1988).

$$Rof = (P - I)^2 / (P - I + S)$$

Where I = Initial retention volume

S = Potential maximum surface retention

The initial abstraction I is considered as 0.2 S

The retention volume is given as

$$S = 25400/CN - 254$$

where, CN is the runoff Curve Number, a parameter dependent on soil type, land use, and antecedent moisture condition. Considering the land use pattern a representative weighted average curve number for the entire watershed has been worked out as 60 (CNI), 78 (CNII) and 90 (CN III).

Generally the soils of the watershed are in Antecedent Moisture Condition I (AMC I) prior to the monsoon and after initial monsoon rains the antecedent moisture condition may change to AMC II or AMC III depending on the rainfall pattern during the monsoon period till early October.

Actual Soil Evaporation : The loss of water by evaporation from the soil surface is a major component of the annual water balance of the semi-arid tropics. The actual evaporation from the soil surface nearly equals the potential evaporation when the soil surface is saturated with water. The maximum depth of the soil where soil evaporation will occur depends on the texture of the soil and it is very difficult to measure in the field. The daily actual soil evaporation AE has been estimated as a function of the daily pan evaporation value Ep, the number of days, t, following the rain of sufficient amount to recharge 20 cm thickness of soil zone from surface and fraction B of incoming solar radiation reaching the soil surface (Russel, 1989). The following equation

$$AE = B. Ep / t$$

is used to compute the daily actual soil evaporation from the daily rainfall data and pan evaporation data combinedly. Under uncropped conditions of barren land B = 1.0, but under cropped conditions it is a time dependent function of crop growth that can be measured directly or estimated from the leaf area index (LAI). Also it was found at ICRISAT Campus that the actual evaporation from soil may not be greater than half of the pan evaporation after applying corrections to pan coefficient (Pathak et al, 1989).

Actual Transpiration : To compute ET crop, a three stage procedure has been proposed by Food and Agriculture Organization (FAO, 1977). The effect of crop characteristics on crop water requirements is given by the crop coefficient (Kc) which represents the relationship between the reference (ETc) and crop transpiration (ET crop)

$$ETcrop = Kc . ETc$$

values of Kc are dependent on the crop, its stage of growth, growing season and the weather conditions.

To convert pan evaporation (Epan) into reference crop transpiration (ETc), emperically derived coefficient (Kp) is given which takes into account climate and pan environment. Reference crop transpiration (ETc) can be obtained from

$$ETc = Kp . Epan$$

where, Epan = Pan evaporation in mm/day and represents the mean daily value of the period considered

Kp = Pan coefficient

Considering average daily pan evaporation at Parkal on yearly basis and the rainfall pattern during monsoon season, the pan coefficient Kp has been selected as 0.8 for moderate wind and medium humidity conditions.

Soil Moisture : Some surface runoff will be generated after a sufficient rainfall event and the remaining rainfall tries to saturate the soil zone upto field capacity and the surplus water if any, leaves the soil zone as groundwater recharge to the underlying water table. The moisture which has remained in the soil zone as the available soil moisture will be lost either as soil evaporation or transpiration by vegetation. A two layer soil zone has been assumed in the water balance model. In the top layer, which is assumed to be upto 20 cm depth from the ground surface both soil evaporation and transpiration could occur, whereas in the second layer, underneath the first layer upto 45 cm from surface, only transpiration by plants could take place as long as the soil moisture is available. The average available moisture holding capacity of the Alfisols of the region ranges from 45-75 mm/45 cm of soil zone (Randhawa and Singh, 1988). A representative average thickness of 45 cm of soil zone has been assumed to be possessing an average water holding capacity of 60 mm.

The component of groundwater recharge could ultimately be obtained from the water balance computation. Various components of the recharge process model, viz., surface runoff, actual soil evaporation, actual transpiration, soil moisture status, and groundwater recharge have been computed following a daily soil moisture accounting procedure.

Results of Recharge Process Model : The annual ground water recharge and other components of the water balance model during 1976-1990 has been shown in Table 2. Groundwater recharge in the watershed mostly takes place during July to September (Table 3).

Table 2 : Annual water balance components in Parkal watershed in mm.

Year	Rainfall	Interception Loss	Surface Runoff	Soil Evaporation	Transpiration	Recharge
1976	1061.4	15.4	352.7	144.2	314.9	234.2
1977	724.6	15.4	57.4	205.8	434.3	11.5
1978	1394.0	18.6	373.5	252.4	540.6	209.2
1979	784.8	15.3	189.2	133.6	341.3	98.7
1980	710.0	15.4	82.8	144.5	381.3	92.6
1981	1144.2	19.8	281.5	191.1	452.3	199.7
1982	1052.0	15.0	205.3	197.6	455.9	177.9
1983	1386.4	14.0	505.7	138.5	359.5	357.5
1984	946.3	14.5	240.6	182.1	426.2	94.3
1985	719.2	14.0	142.2	145.8	359.6	57.7
1986	1087.0	12.5	411.7	137.9	419.5	91.5
1987	873.7	13.4	199.2	212.0	414.6	36.9
1988	1501.4	12.0	549.6	214.4	443.9	281.4
1989	1346.6	7.8	527.5	144.4	394.5	272.4
1990	1628.1	12.5	465.8	262.5	645.3	241.9
AVG.	1090.6	14.4	305.6	180.4	425.6	163.8

Table 3 : Monthly Groundwater Recharge During 1981–1990 in mm

S. No.	Year	Jan.	Jul.	Aug.	Sep.	Oct.
1.	1981	16	57	96	20	11
2.	1982	–	72	106	–	–
3.	1983	3	44	125	108	76
4.	1984	–	37	23	26	9
5.	1985	–	3	54	–	–
6.	1986	–	13	77	2	–
7.	1987	–	37	–	–	–
8.	1988	–	152	55	75	–
9.	1989	1	239	32	–	–
10.	1990	92	32	72	–	22

The average annual rainfall, the estimated surface runoff and groundwater recharge from the water balance model of the watershed are 1090 mm, 305 mm and 164 mm respectively under normal agroclimatic conditions. The percentage of mean annual groundwater recharge and surface runoff with respect to annual rainfall works out to be 15 and 28 respectively during 1976-1990. The percentage of the average annual recharge of 15 with respect to the annual rainfall for the Parkal watershed is comparable with the recharge rates of 14.5% of annual rainfall obtained for other granitic watershed viz., Dulapally near Hyderabad (Narasimha Reddy et al, 1991). The recharge estimates of Vadavathi river basin in parts of Karnataka and Andhra Pradesh on a similar granitic terrain are reported to be varying between 13-20 percent of annual rainfall (Sukhija and Rao, 1983).

B. AQUIFER MODEL

The computer code NEWSAM (Ledoux et al., 1989) solves the system of groundwater flow equations using integrated finite difference method by implicit discretization of time steps. The linear system is solved iteratively by the point successive over-relaxation method. Figure 1 shows the nested squares meshes of Parkal watershed aquifer model. A well distributed 13 representative observation wells have been selected for construction of well hydrographs from June, 1981 to May, 1991. The water level configuration of June, 1981 has been assumed to be in equilibrium condition and therefore, has been considered as initial water level for aquifer modeling. The water levels between observation wells have been interpolated taking care of surface topography, stream bed elevations and the water level in the surface water bodies. Elevations of beds on surface water bodies have been interpolated to infer stream bed elevations and have been simulated accordingly as known water levels in the model.

Aquifer parameters viz., transmissivity and storage coefficient were estimated by conducting pumping tests on four dug wells in the area. Transmissivity value are varying from 14 to 42 m^2/day (Fig. 2) and the average specific yield is 0.054. The transmissivity values have been assigned to the corresponding nested square meshes and at the remaining meshes the 'T' values have been assigned through interpolation considering geomorphologic and subsurface geological features. The boundaries have been realized by terminating the meshes with no flow boundary condition by assigning zero transmissivity values in the model.

Recharge to the groundwater regime due to monsoon rainfall forms the main input to the aquifer system. Seepage from surface water bodies and irrigation return seepage from paddy fields also contributed as input stresses to the flow regime. The outflow is mainly through groundwater withdrawal from open wells and borewells during non-monsoon season mainly for irrigation and intermittent base flow towards streams during monsoon season. The base flow joins the outflow and leaves the watershed at Paidpally tank outlet.

Steady State Calibration : The groundwater draft has been estimated based on well inventory and average running hours of pumping and the cropping pattern, and unit draft of 0.67 ham/annum has been considered initially. Draft values have been assigned to each mesh based on density of wells falling in a particular mesh. The average groundwater recharge estimated from the water balance model (164 mm/per annum) has been uniformly fed at all the meshes, except at those meshes falling on the stream courses and surface water bodies. The seepage from surface water bodies through beds of the tanks at a rate of 46 cm/year is distributed appropriately over 9 months in a year in the model.

The transmissivity values modified during the steady state model calibration are varying from 15 to 45 m^2/day which are comparable with the estimated transmissivities from pumping tests. The computed water levels in the steady state calibrated model are found to be matching with the observed water levels within ± 1.0 m(Fig. 3). It was noticed during the processes of model fitting that the variation of transmissivities produced negligible changes in the computed water levels and only the input and output stresses determined the water level configuration.

The water balance of the groundwater regime of Parkal watershed for steady state condition is summarised in Table 4. An average annual input stress 2.18 MCM (Million Cubic Metres) consists of recharge due to rainfall, seepage from surface water bodies and irrigation return flow from paddy fields under tank ayacut areas. The output stresses consist of groundwater pumpage from open wells and borewells to the tune of 1.67 MCM (Fig. 4) a base flow of 0.27 MCM towards streams and a subsurface outflow of 0.24 MCM across the Paidpally tank bund.

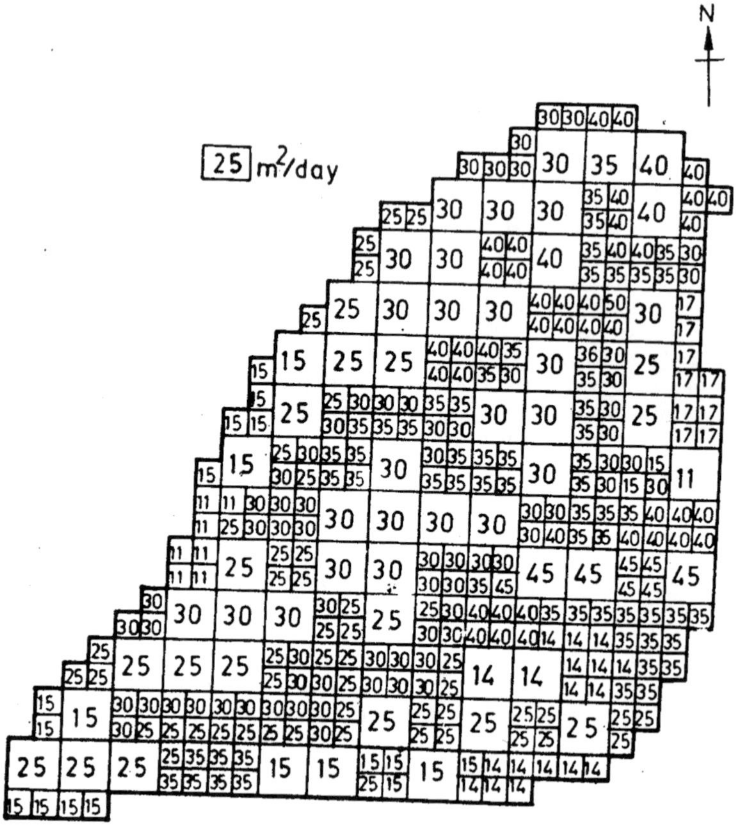

Fig. 2. Modified Spatial distribution of Transmissivity (Calibrated Model)

Table 4 : Average Annual input and output stresses for steady state (June, 1981) (million cubic meters) MCM

RUN	INPUT	OUTPUT	
	Recharge due to rainfall seepage from tanks and irrigation return	Ground water draft	Base flow to streams and outflow
1.	2.18	1.80	0.38
2.	2.18	1.61	0.57
3.	2.18	1.67	0.51

Transient State Calibration : The average specific yield (Fig. 5) determined through pump tests have been assigned at all the mesh nodes. The dynamic variation of draft as well as the groundwater

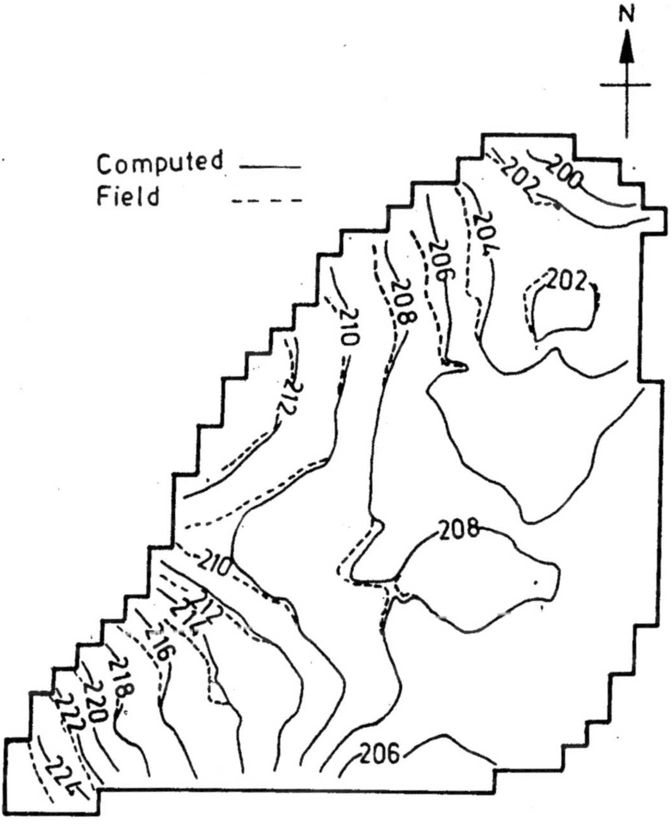

Fig. 3. Comparison of computed and observed water level contours in m (AMSL) - Steady State (June 1981) Calibrated model.

recharge estimated earlier has been fed to the aquifer model in monthly time steps during transient condition (Table 3). The groundwater flow model has been calibrated for 10 year period from June 1981 to May 1991 through comparison of computed and field well hydrographs at the observation wells. The time- variant draft of 1.5 MCM/annum was maintained till 1986-87 and later on a draft of 1.6 MCM/annum, which is 6 percent more, has been given in the model. Some relative reduction of annual draft has been effected during good rainfall years using monthly rainfall information and availability of surface water in the tanks.

Table 5 : Annual input and output stresses for transient condition in MCM

SL. No.	Year	Input: Recharge due to Rain and Irrigation return	Input: Ground Water Draft	Output: Baseflow to Streams & Outflow	Output: Change in storage
1.	1981-82	2.50	1.50	1.00	0.00
2.	1982-83	2.39	1.54	0.86	-0.01
3.	1983-84	4.31	1.51	1.60	+1.20
4.	1984-85	1.40	1.45	0.86	-0.91
5.	1985-86	1.00	1.58	0.30	-0.88
6.	1986-87	1.30	1.57	0.13	-0.40
7.	1987-88	1.20	1.55	0.00	-0.35
8.	1988-89	3.50	1.60	0.81	+1.06
9.	1989-90	4.80	1.60	1.42	+1.78
10.	1990-91	1.70	1.50	1.15	-0.95

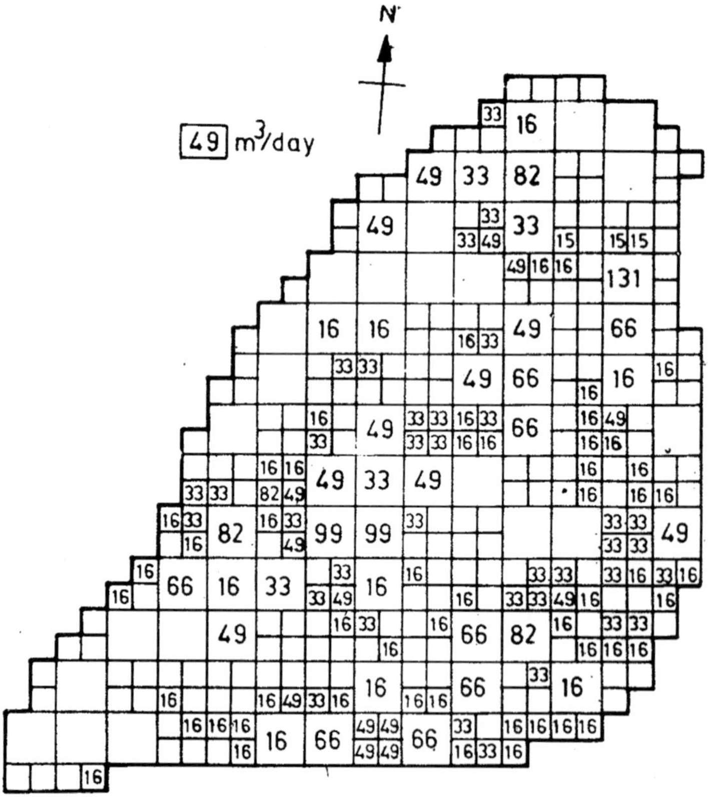

Fig. 4. Modified Spatial distribution of draft (1981)

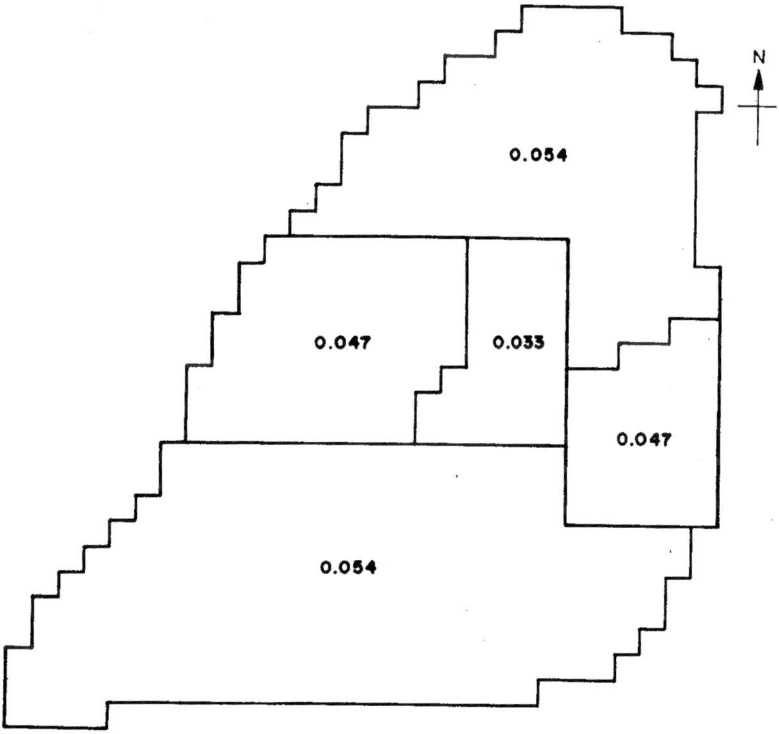

Fig. 5. Regional Specific yield Values (Calibrated model)

The input and output stresses used for the transient simulation are shown in Table 5. The draft has been redistributed at some meshes and the specific yield values have been localised and assigned to meshes to obtain a close match between computed and observed well hydrographs particularly at well no. 16B, 33 and 1. At this stage the computed and observed well hydrographs are found to be not matching during 1986 and 1987 at most of the observation wells. This may be attributable to some extraordinary recharge occurring during August 1986 and it could not be simulated in the model due to lack of understanding on the preferred path ways flow, which is happening after two years of drought period. Hence, some additional recharge has been given during August 1986 to bring the computed well hydrographs during 1986 closer to the observed one in the transient model. The comparison of computed and observed well hydrographs at the observation wells is shown in Figure 6 (A,B,C)

CONCLUSION

The prognostic capability of deterministic groundwater flow models allows them to be used as a basis for evaluating options for managing groundwater resources. Management models for groundwater flow were initially developed in response to the growing need to understand systematically the hydraulic behavior of the flow system to economize the utility of the scarce groundwater supplies. Aquifer management research also envisages the problems of groundwater pollution. The predictive accuracy of the model does not necessarily represent its basic value. Rather it provides a means of quantitatively assess and assure the consistency within and between concepts of the relevant coefficients. In this way, a simulation model helps the investigator to improve the understanding of the various factors controlling groundwater flow.

Presently the groundwater modeling has evolved to a level at which it has become a worthwhile tool for the hydrogeologist tackling the complex problems encountered in day to day work. However, model use should be tempered by an adequate awareness of the limitations, underlying assumptions and appropriateness of the model. The final worth of the modeling applications ultimately depends on the people who apply the models, their knowledge of the site hydrology, and their understanding of modeling capabilities. The computational errors associated with or produced by (a) theoretical misconceptions or over idealizations about the system that are incorporated into the model, (b) uncertainity and error in the specification of system properties, boundary conditions, and initial conditions, and (c) uncertainity in past and future stresses. Presently the groundwater flow modeling has reached a stage that equation solution and simulation code are no longer a priority research area. Instead, much current research aims on parameter uncertainity, groundwater management models, parameter estimation, and flow in fractured media.

An aquifer simulation model is not mere an approximation of a complex field situation. Improvements in the approximation are always possible, thus, models should be viewed as dynamic representations of nature, subject to further refinement and improvement. As new information becomes available, previous forecasts could and should be modified. Feedback from preliminary models not only helps the modeler to set improved priorities for the collection of additional data, but also helps test hypothesis concerning governing processes in order to develop an improved conceptual model of the system and problem of

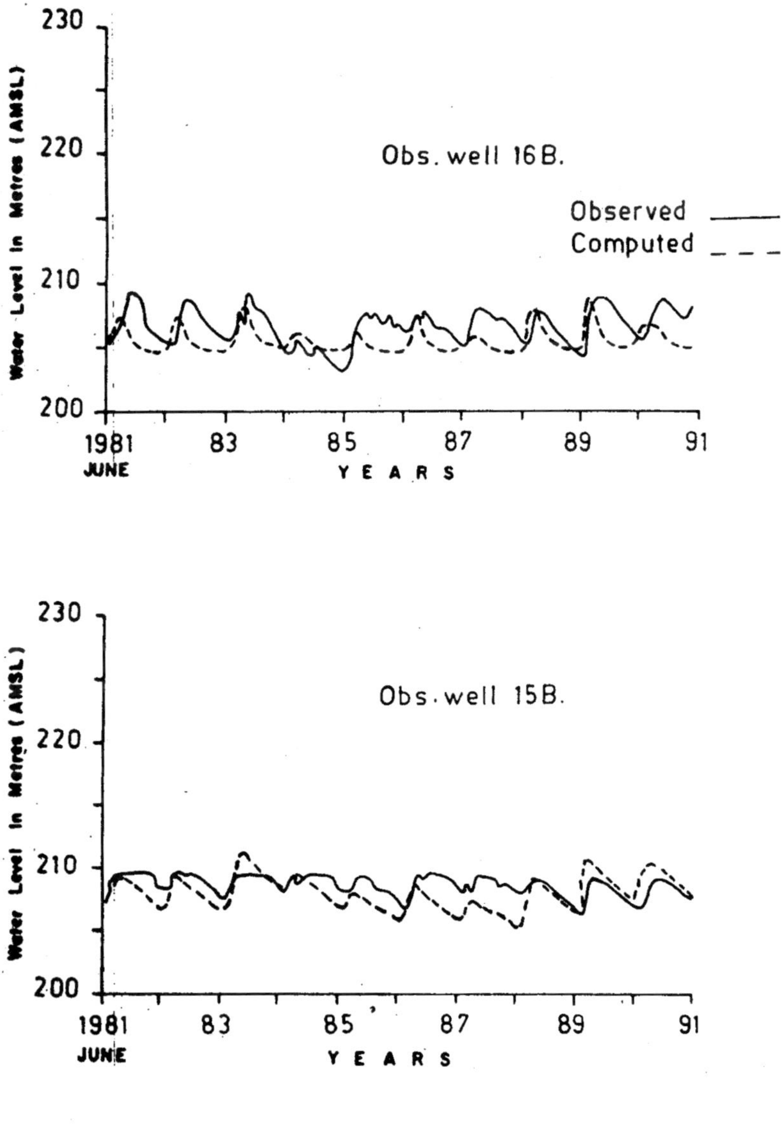

Fig. 6A. Comparison of Computed and observed hydrographs Transient Condition - (Obs. well 16B and 15B)

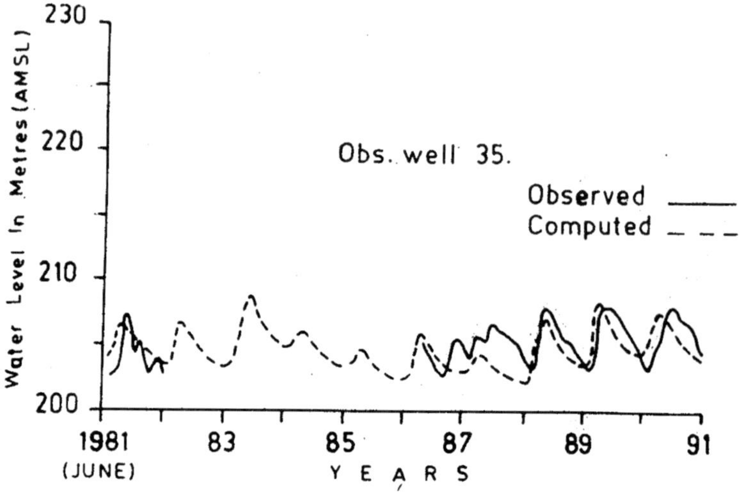

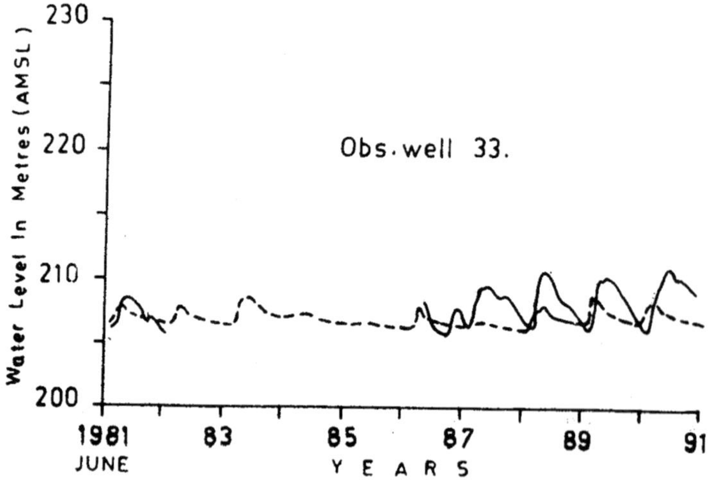

Fig. 6B. Comparison of Computed and observed hydrographs Transient Condition - (Obs. well 35 and 33)

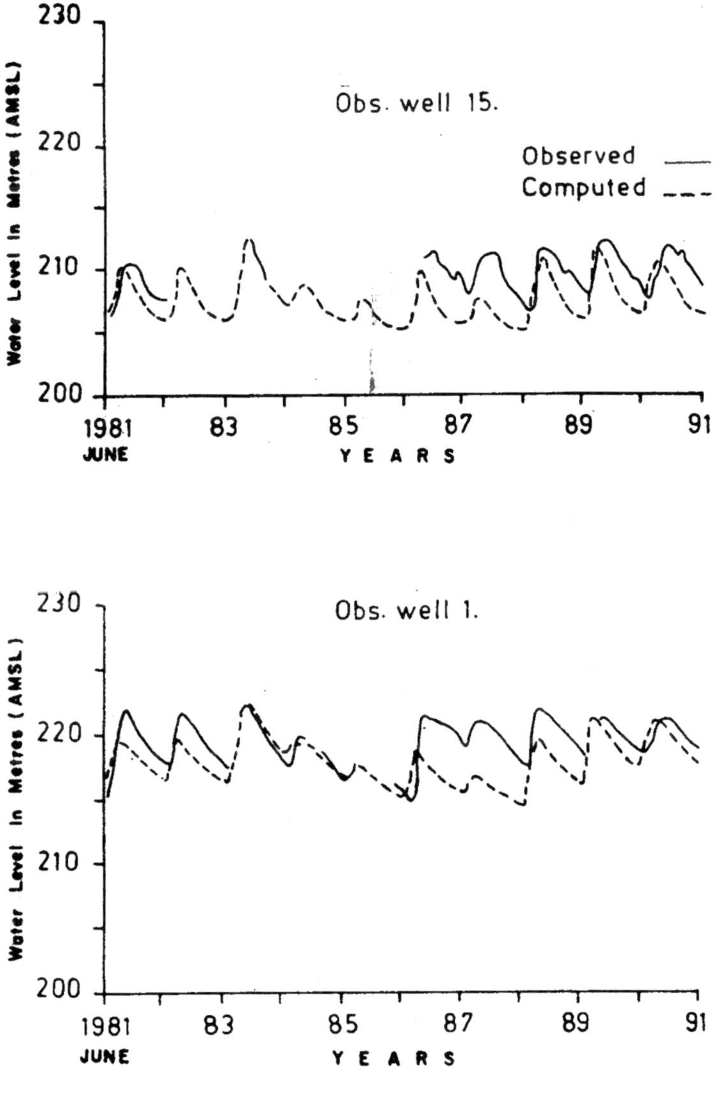

Fig. 6C. Comparison of Computed and observed hydrographs Transient Condition - (Obs. well 15 and 1)

concern. In summary, the primary value in many cases of deterministic groundwater model analyses is in providing a disciplined format to improve one's understanding of the aquifer system.

REFERENCES

Bear, J. 1979. Hydraulics of Groundwater. Mc.Graw Hill, New York, NY, 567 pp.

De Marsily, G. 1986. Guantitative Hydrogeology. Academic Press, New York, Ny, 440 pp.

Delhome, J.P. 1979. Spatial variability and uncertainity in groundwater flow parameters : A geostatistical approach., Water Resour. Res. 15(2) : 269-280.

De Marsilv, G. Ledoux, E., Leavassor, A., Poitninal, D., and Salem, A., 1978. Modeling of large multilavered aquifer systems : Theory and Applications. *Journal of Hydrology*, 36, 1-34.

Doorenbos, J. and Pruitt, W.O., 1977. Guide lines for prediciting crop-water requirements. FAO Irrigation and Drainage Parper 24 (Revised), FAO, Rome, pp. 144.

Faust, Charles R, Silk Lyle,R. and Mercer, J.W. 1981. Computer Modeling and groundwater protection. Groundwater Vol. 19, No. 4, pp. 362-365.

Freeze, R.A. 1982. Hydrogeological concepts in stochastic and deterministic rainfall-runoff predictions. In : T.N. Narasimhan (Editor), Recent Trends in Hydrogeology. Geol. Soc. Am. Sec. Pap. 189 : 63-79.

Gorelick, S.M. 1983. A review of distributed parameter groundwater management methods. Water Resour. Res, 19(2) : 305- 319.

Gorelick, S.M., Voss, C.I., Gill, P.E., Murray. W., Saunder, M.A., and Wright, M.H. 1984. Aquifer reclamation design : The use of contaminant transport simulation coupled with non-liner programming. Water Resour. Res. 20(4) : 415-427.

Groundwater Department, A.P., 1976. Interim Report on studies in Prakal Critical basin, Partkal taluk. Warangal district with Special Refrence to optimum density of wells. Miscellaneous Report no. 67.

Gupta, C.P., Thangarajan, M., Gurunadha Rao, V.V.S., Balasundar, N.K. and Muthuram, S. 1989. Simulation of regional hydrodynamics in the Neyveli aquifer. In the proceedings of International workshop on Appropriate Methodologies for Development and Management of Groundwater Resources in Developing Countries held at NGRI during Feb 28-Mar 4. pp. 607-618.

Javandel, I., Doughty, D. and Tsang, C.F. 1984. Groundwater transport : Handbook of mathematical models. Am. Geophys. Union, Water Res. Monogr, 10 : 228 pp.

Konikow, L.F., and Mercer, J.W. 1988. Groundwater flow and Transport Modeling. Jour. Hydrology, Vol 100, No. 2, pp. 221-270.

Konikow, L.F., and Patten, E.P., Jr 1985. Groundwater forecasting in M.G. Ander son, and T.P. Bent (Eds.) Hydrolosical forcasting, wiley, New York PP. 221.-270.

Manteuffel, T.A. Grove, D.B. and Konikow, L.F. 1983. Application of the conjugate gradient methods to groundwater models. U.S. Geol. Surv., Water Res. Invest. Rep. 83-4009, 24pp.

Mercer, J.W. and Faust, C.R. 1981. Groundwater modeling. National Water Well Associ., Worthington. Ohio, 60pp.

Moore, J.E. 1979. Contribution of groundwater modeling to planning. Journ. of Hydrology. V. 43 (Oct)pp. 121-128.

Narasimha Reddy, T., Prakasam, P., Nageswara Rao T., Subrahmanyam, G.V., Prakash Goud, P.V., Harish, K., Gurunadha Rao, V.V.S. and Gupta, C.P., 1991. *Water balance model and groundwater flow model in Dulapally basin*, Groundwater Department, A.P., Tech. Report Series 9.

Narasimha Reddy, T. Prakasam, P. Subrahmanyam, G.V. Prakash Goud, P.V. Gurunadha Rao, V.V.S. and Gupta, C.P. 1992. Characterization of groundwater recharge in a crystalline watershed through recharge process and groundwater flow models, Seminar on Groundwater Hydrology, New Delhi. pp. 7-21.

Pathak, P., Laryea, K.B. and Sudi, R., 1989. A runoff model for small watersheds in the semi-arid tropics, *Tanasactions of the ASCE*, Vol. 32 no. 5, p. 1619-1624.

Randhawa, N.S. and Singh, R.P., 1988. Fertilizer management in rainfed areas Available technologies and future needs. *Fertilizer News*, 28(9), p. 17-32.

Remson, I., Hornberger, G.M. and Molz, F.J. 1971., Numerical methods in sub-surface hydrology, Wiley, New York, NY, 389pp.

Russel, M.B., 1989. Profile moisture Dynamics of soil in Vertisols and Alfisols. *Proceedings of the International Workshop on the Agroclimatological Research needs of the Semi- Arid Tropics*, 22-23 Nov., 1978, ICRISAT, Patancheru, 502-324, India.

Sukhija, B.S. and Rao, A.A., 1983. Environmental trittium and radio-carbon studies in the Vedavati River basin, Karnataka and Andhra Pradesh, *India, J. Hydrol*, 60, p. 185-196.

Ven Te Chow, D.R. Maicment and Lary W. Mays, 1988. *Applied Hydrology*, McGraw-Hill International Editions, Civil Engineering Series, p. 147-150.

12

WATER POLLUTION AND LEGISLATION

K. V. RAMANA MURTHY

Pendekanti Law College, Hyderabad-500 029

Pollution of water is now posing a serious threat in many regions in India. Unprecedented population growth, rapid pace of industrialization and unplanned urbanization are primarily responsible for causing pollution in rivers, streams, lakes and other surface water resources. Even the prestine groundwater resources are not devoid of this pollution threat.[1] In some ways the pollution situation in India is worse than in some of the industrialized countries of Western Europe and America.[2] In India, for instance, almost 70% of the available fresh water is feared contaminated by pollutants.[3] According to the World Health Organization 80% of the diseases in the world are imputable to inadequate water and the related problem of sanitation. It has been stated that "two-thirds of all illness in India is related to water borne diseases such as typhoid, infective hepatitis (jaundice), cholera, diarrhoea, and dysentry. Many of them assume epidemic proportions".[4] Lack of proper water treatment or sewage disposal facilities[5] in most of the towns and cities have greatly contributed in polluting the water resources. Ironically, water pollution did not receive required attention of planners until recent times. The awareness for evolving a comprehensive regulatory mechanism for the control of water pollution and for the preservation and conservation of our water resources is still growing.

Prevention and control of water pollution form an integral part of environmental protection. It requires a closely co-ordinated series of legislative, administrative and technical measures. Water pollution problems are addressed in India by various legislations.

GROUND WATER

Editor: P. Sankara Pitchaiah, © 1994 Scientific Publishers, Jodhpur

EARLY LEGISLATION BEARING ON WATER POLLUTION

Indian Penal Code, 1860

Causing defilement of water in public spring or reservoir is an offence under Indian Penal Code.[6] The rationale behind this section is that the water of a public spring or reservoir is a common property. Therefore, no one should disturb its public enjoyment by an act inconsistent with its beneficial use. To constitute an offence under this provision there must be some physical act of defiling water. An act is actionable only when such act is voluntary and it renders the water of a public spring or reservoir less fit for the purpose for which the water is ordinarily used. This provision, however, is very limited in its scope. It applies only to "Public spring or reservoir", but does not cover "continuous stream of water running along the river bed".[7] Section 290 of the Code attempts to rectify this lacuna. It reads :

> "Whoever commits a public nuisance in any case not otherwise punishable by this code, shall be punished with fine which may extend to two hundred rupees".

As contaminating river water and rendering it unfit for drinking purposes would amount to committing public nuisance[8] it should be punished under this section.

An instance of causing water pollution can also be dealt with as 'mischief'[2] under the penal code. Mischief is defined as :

> "Whoever with intent to cause or knowing that he is likely to cause, wrongful loss or damage to the public or to any person, causes the destruction of any property or any such change in any property or in the situation thereof as destroys or diminishes its value or utility or affects injuriously commits mischief".[9]

In order to constitute the offence of mischief it is necessary to have some direct act on the part of the mischief-doer either personally or through someone else. Intention or knowledge is the basis for imposing the criminal liability under this provision.

However, these provisions of the Code are very outdated in the present context to effectively deal with modern problems of water pollution for the punishment prescribed for the commission of the said offences is meagre and they fail to provide deterrent effect.[10]

Indian Easements Act, 1882

Easements Act contains provisions for the protection of riparian rights. A riparian owner is one who has title to land adjacent to a natural stream. Under section 7 of the Act every riparian owner has the right to the continued flow of the waters of a natural stream which passes by, through or over his land in its natural condition without obstruction or unreasonable pollution.[11] However, regarding groundwater the common law makes a distinction between underground percolating water and underground stream. Underground stream means water running underground in defined or ascertained course. Underground percolating water does not flow in defined channel. As regards water flowing underground in a defined channels land owner's rights[12] are limited by corresponding rights of other riparian rights. Every riparian owner has a right to the accustomed flow of water without material alteration in quantity, direction, force or temperature.[13] No such limitation is placed in respect of groundwater running in undefined and unknown channels. The arguments advanced for the acceptance of this rule was that the source and flow of these waters were so unknown[14] and that it is impossible to formulate any legal rules[15] governing them. It suggests that a person by abstracting percolating groundwater may even drain water to such an extent that his neighbour may be left without any percolating groundwater even though the latter was the prior user of such water. The owner by such tapping is said not to infringe any right of his neighbour even if the effect of his action is to affect the quality of water in the neighbour's land.[16] This common law doctrine of land ownership which is adopted under the Easements Act is playing havoc in the coastal regions. In coastal regions fresh groundwater supplies are susceptible to contamination by salt water intrusion. Invasion of saline water into fresh groundwater due to excessive withdrawal of groundwater commonly occurs in coastal aquifers.[17]

Attakoya Thangal vs. *Union of India* [18] was an illustrative case on this point. Petitioners in this case approached the High Court seeking an appropriate direction restraining the Lakshadweep administration from implementing a scheme to augment fresh water supply by digging wells and by using electric or mechanical pumps. In the opinion of the petitioners, large scale withdrawal of groundwater can deplete the water resources causing seepage or intrusion of salt water from the surrounding Arabian sea. Allowing the petition the court observed that right to sweet water and right to free air are attributes of the right to life guaranteed under Article 21 of the Indian Constitution for they being the basic elements to sustain life.[19] The court also directed the administration to make a deeper study to know whether the process would not dry up or result in salt water intrusion into the coastal aquifers.[20] Highlighting the immediate need for evolving safeguards to stop withdrawal of groundwater at a cut off level and introducing a system of effective monitoring at all levels, the court recommended for a comprehensive statutory regulation to protect groundwater resources in the coastal regions.[21]

Code of Civil Procedure, 1908

Public nuisance action can be brought before a court either by a civil or criminal action. Section 91[22] of the Code of Civil Procedure, 1908 ensures the right of action in the case of public nuisance. In England, only an Attorney General or Local Prosecutor is permitted to sue and abate public nuisance.[23] A private individual may bring a private action on public nuisance in his name only when he can show that he has suffered some particular foreseeable and substantial damage over and above that sustained by public at large or when the interference with the public right involves a violation of some private right of his own. But in India, legislation regarding public nuisance went a step forward and it enabled two or more persons, with the leave of the court to institute a suit for declaration or injunction or for such relief as may be appropriate in the circumstances of the case, even though no special damage has been caused to them by reason of such public nuisance.[24]

Code of Criminal Procedure, 1973

The Code of Criminal Procedure contains a significant provision which gives preventive as well as remedial jurisdiction to the Magistrate to deal with Public nuisance.[25] The exercise of this power for removal of public nuisance was at stake before the Supreme Court in *Municipal Council, Ratlam.* v. *Vardhichand.*[26] In that case the residents of locality of Ratlam Municipality tormented by stench and stink caused by open drains and public excretion by nearby slum dwellers moved the Magistrate under section 133 of the Criminal Procedure Code to require Municipality to do its duty towards the members of the public by removing the existing nuisance. The municipality expressed its financial inability to carry out its duties. Thus the question before the Supreme Court was whether the financial inability of a municipality exonerates it from its statutory liability to do certain public duties. Commenting upon the obligation of the municipality to abate public nuisance, the Supreme Court ruled that the municipality cannot extricate itself from its responsibility. In this context the court observed:

> "The Criminal Procedure code operates against statutory bodies and others regardless of the cash in their coffers even as human rights under Part-III of the Constitution have to be respected by the State regardless of budgetary provision.[27]

The Court regarded public nuisance as a challenge to social justice component of rule of law.[28] Decency and dignity are non- negotiable facets of human rights and are a first charge on local self governing bodies. The public power under section 133 is a public duty to the members of the public who are victims of public nuisance.[24]

The Supreme Court had taken a similar view in *Gobind Singh* v. *Shanti Sarup* [30] regarding the health, safety and convenience of the public at large.

A subsequent decision of the Madhya Pradesh High Court in *Krishna Gopal* v. *State of M.P.* [31] demonstrates the trend set by *Ratlam Municipality case*. A glucose saline manufacturing unit, which was licensed ignoring the objections raised by the local residents, caused noise and air pollution disturbing the sleep of the heart patient living nearby. On complaint made by the patient's wife the District Magistrate issued orders directing the removal of the factory. An appeal to High Court was preferred by the factory management against the directions of the Magistrate. The contention of the factory management was that the alleged act amounted only to private nuisance and not public nuisance. It was not accepted as the police report proved contrary. In the opinion of the Court whether or not it affects one individual or more, hazardous activity having the potential to cause loss or injury to more than one individual is to be considered as public nuisance and section 133 can be invoked when the jurisdictional facts are present.

However, in 1989 the same High Court in *Abdul Hamid* v. *Gwalior Rayon Silk Manufacturing Co. Ltd.*, [32] refused to invoke section 133 of the Code of Criminal procedure when jurisdictional facts existed requiring an immediate action. It was a case where the appellant alleged that respondents were responsible for the large scale pollution of water resulting from the disposal of industrial wastes. The Court found that the courts cannot entertain section 133 action to abate water pollution. It reasoned that the pollution laws are there to ensure proper balance between the conflicting demands of nation's industrial progress and health hazards to the people. According to the Court all the remedies which could be provided by the Magistrate under section 133 could certainly be provided under Water Act, 1974, a comprehensive law dealing with prevention and control of water pollution. Pollution control boards conceived under this Act, having considerable expertise and requisite machinery can certainly be expected to move purposefully and fruitfully in the case of water pollution. The court categorically ruled that the provisions of section 133 of the code in so far as they relate to water pollution must be deemed to have been impliedly repealed.

Considering the scheme envisaged under section 133 of the Criminal Procedure Code and separate law dealing with water pollution it appears that the view taken in *Abdul Hamid* case is correct. The whole texture of Water Act is designed in such a manner that the authorities established under the Act would have complete control over its administration and prosecution of offenders.

Local Bodies Legislation

Local Bodies have to play a significant and decisive role in the cause of environmental protection by taking effective measures for the protection and maintenance of the quality of water within their respective jurisdictions. Almost all state governments in India have enacted legislations imposing responsibilities on the local authorities to look after the environmental needs of their territories. They have the duty to construct public conveniences and keep them clean and hygenic, provide facilities for the disposal of rubbish, filth and carcases, see the industrial activities do not result in unreasonable pollution and take measures to protect the health of the people from dust, smoke and contaminated water.

Post-Stockholm Era

Ever since the United Nations Conference on Human Environment held at Stockholm in 1972 proclaimed the right of people to a clean and healthy environment, the Government of India as well as the State Governments have initiated a number of measures to control various types of pollution.

Constitutional Developments

India is one of the few countries in the world that have made reference in the Constitution to the need for the protection of the environment. Within five years from 1972 conference, the Constitution of India was amended *inter alia* to include environmental protection as a constitutional mandate. By the Constitution (Forty-second Amendment) Act of 1976 protection and improvement of environment and safeguarding the forests and wildlife were declared as directive principles of state policy which shall guide the legislative and administrative endeavours.[33] Under this provision the state may not only adopt protectionist policy but also provide for the improvement of polluted environment. Besides protection of natural environment including forests, lakes, rivers and wildlife and compassion for living creatures were made the fundamental duty of every citizen.[34]

The role of Indian judiciary especially in the recent years has been one of dynamic activism. The courts are innovating new methods and devising new strategies for providing access to justice. The judicial system established under the constitution has proved that it has been a success in safeguarding the right to life, liberty and pursuit of happiness. Emphasising the need for pollution free environment, the Court in *Rural Litigation Entitlement Kendra, Dehradun* v. *State of U.P.*[35] Observed that the restriction imposed on the mining operations must be considered as the price to be paid for the protection of the right of the people to live in a healthy environment. Though the court did not overtly state that right to clean environment is envisaged under Article 21 of the Indian Constitution which guarantees right to life, the fact of accepting the petition of under Article 32 indicates the judicial recognition of such right. Nevertheless, the High Courts are deliberate and categoric in reading, the right to clean environment into fundamental right to life. In *Damodar Rao* vs. *The Special Officer, Municipal Corporation of Hyderabad* [36] the Andhra Pradesh High Court held that the acquisition of a part of land which was previously earmarked under the development plan for construction of housing colony was contrary to the vigorously growing law of environment, and the right to enjoy life guaranteed under the Constitution. The court observed :

> ..... the enjoyment of life and its attainment and fulfilment guaranteed by Article 21 of the Constitution embraces, the protection and preservation of nature's gifts without which life can not be enjoyed. There can be no reason, why practice of violent extinguishment of life alone should be regarded as violative of Article 21 of the Constitution. The slow poisoning by environmental pollution and spoilation should be regarded as amounting to violative of Article 21 of the Constitution." [37]

Similar view was taken in *Attakoya Thangal* vs. *Union of India* [38] by the Kerala High Court wherein it observed that right to sweet water and free air are attributes of right to life as they being the basic elements to sustain life.

Water (Prevention and Control of Pollution) Act, 1974

As early as 1962, the Ministry of Health had appointed an expert committee on water pollution. The committee recommended that Union as well as State laws on the subject be enacted. In 1963, the General Council on Local Self Government recommended the enactment of a single law by Parliament. Consequently draft bill was circulated to all State Governments in 1965, with a request to pass enabling resolutions as required in Article 252 (1) authorising the Parliament to enact such a law on their behalf.

The prevention of Water Pollution Bill, 1969 was introduced in the Rajya Sabha after six States had adopted the required enabling resolutions. In August 1970, the Rajya Sabha decided to refer the Bill to a joint committee which modified it in many respects and then presented it to the Parliament in 1972. The Bill was passed by the Parliament in 1974, which took the form of the Water (Prevention and Control of Pollution) Act 1974 (hereinafter, referred to as the Water Act).

The Water Act is the statute par excellence on water pollution control in India. It seeks to maintain or restore "wholesomeness of water". It seeks to ensure that the domestic and industrial effluents are not allowed to be discharged into water courses without adequate treatment. The Act is quite comprehensive in its coverage of various types of water. For the purpose of the Act a "stream" includes river, water course (whether flowing or for the time being dry), inland water (whether natural or artificial), subterranean waters and sea or tidal waters to the extent specified in notification in the Official Gazette.[39]

Pollution is defined in the Act as "such contamination of water or such alteration of the physical, chemical or biological properties of water or such discharge of any sewage or trade effluent or of any other liquid, gaseous or solid substances into water (whether directly or indirectly) as may, or is likely to create nuisance or render such water harmful or injurious to public health, or safety or to domestic, commercial, industrial, agricultural or other legitimate uses, or to the life and health of animals or plants or of aquatic organisms.[40] This definition displays an awareness of the manifold adverse effects of water pollution and the need to adopt a general approach to the problem.

The Water Act creates agencies for pollution control. It provides for the establishment of Central Board and State Board for states, for prevention and control of water pollution.[41] The functions of these Boards are of deliberative, advisory, investigating and research oriented in character.[42] Functions of these boards include :

(i) Evolving and planning a comprehensive programme for prevention, control and abatement of water pollution;

(ii) Conducting, encouraging, and participating in investigations and research of water pollution problems;

(iii) Inspecting facilities for sewage and trade effluents treatment and

(iv) Developing economical and reliable methods of treatment of sewage and trade effluents.

Pollution control boards are empowered to control water pollution primarily by way of standards laid down[43] by them and issue of consent orders.[44] For violating consent orders or the standards laid down stiff penalties have been prescribed under the Act.

Consent Mechanism

Prevention and control of water pollution is done through a system of consent. The Act generally prohibits disposal of polluting matter into streams, wells and sewers or on land in excess of the standards established by the State Board.[45] Every person must obtain consent from the State Board before taking steps to establish any industry, operation or process, any treatment and disposal system or any extension or addition to such a system which might result in the discharge of sewage or trade effluents. The State Board may impose conditions as to the point of discharge, construction and use of outlet composition and volume of discharge. While granting consent the State Board may evolve standards for effluents for each industry taking the quality of receiving waters and others relevant matters into consideration. Thus, it is the responsibility of the polluting industry to set up treatment plant for bringing effluent standards to the required level. A register has to be maintained by the state board containing particulars of the conditions imposed while granting consent, which shall be open to inspection at all reasonable hours by any person interested in the matter. Further, the State Board is empowered upon thirty days notice to the polluter, to execute any work required under the consent order which has not been executed. It may recover the expenses for such work from the polluter. [46]

Stressing the obligation of the industry to establish treatment plant the Supreme Court in *M.C. Mehta* v. *Union of India* [47] observed :

> "The financial capacity of the tanneries should be considered as irrelevant while requiring them to establish primary treatment plants. Just like an industry which can not pay minimum wages to its workers can not be allowed to exist, a tannery which can not set up a primary treatment plant can not be permitted to continue to be in existence for the adverse effect on the public at large which is likely to ensure the tannery to the river Ganga would be immense and it will outweigh any inconvenience that may be caused to the management and labour employed by it on account of its closure. [48]

Preventive and Remedial Measures

Though Water Act states that it is aimed to prevent and control water pollution, a perusal of the enforcement measures show that, preventive aspect has been paid less attention. Under the Act, only in the case of emergency situation, the State Board has the power to take immediate measures to mitigate the harm caused to any stream or well.[49] It is to be noted that the situation where the water has been already polluted is recognised as the emergency situation. In other words, it is only when any poisonous, noxious or polluting matter is present in any stream or well due to any accident or other unforeseen act or event the Board has been empowered to take such remedial measures. The Board shall carry out such operation as may be necessary for any of the following purposes :

(a) Removing that matter from the stream or well and disposing it of in such manner as the Board considers appropriate.

(b) Remedying or mitigating any pollution caused by its presence in the stream or well;

(c) Issuing orders immediately restraining or prohibiting any poisonous, noxious or polluting matter in the stream or well or from making in sanitary use of stream or well.

But it is to be realized that timely preventive measures are more effective than belated remedial action. Boards must have more powers to take necessary action to control the damage at the source itself. But under the Act, the State Board has no power to take immediate action in the case of apprehended pollution. Section 33 of the Act provides for a procedure which has to be complied with in the case of apprehended pollution. The Board can not straightaway interfere to prevent or remedy the situation. The Board has to make an application to the Court not inferior to that of a Metropolitan Magistrate or a Judicial Magistrate of the First Class. It can not initiate any action in the absence of a specific order from the court to that effect. However, it would be difficult for the Board to obtain court orders immediately. The court may initially direct the person who is likely to cause or has caused the pollution to take preventive or remedial measures, as the case may be. It is only after noting the failure of such person in taking measures the Board may be authorised to take necessary action. It is a time consuming process since the Court proceedings involve inordinate delay. Thus, if the Board fails to restrain the person who is likely to cause pollution, intime, one of the fundamental purposes of the enactment, i.e., prevention of pollution would be defeated.

Water Act was amended in 1988 providing for the inclusion of a new section 33-A which empowers the pollution control boards to issue binding directions to any person, officer or authority.

The power to issue directions includes the power to direct (a) the closure, prohibition or regulation of any industry, operation or process; or (b) the stoppage or regulation of the supply of water, electricity or any other service. Pollution control boards, however, could only exercise this power if the pollution arose from any accident or other unforeseen act or event.[50] It will be more functionally dynamic if the boards are given direct control to check apprehended pollution by issuing binding directions, orders or instructions.

Criminal Sanctions

The Water Act employs stringent penal measures for ensuring effective enforcement of its provisions. Establishing or taking steps to establish any industry, operations or process or any treatment and disposal system or bringing into use a new or altered outlet for the discharge of sewage without obtaining the consent of the pollution control boards is an offence under the Act. The minimum punishment prescribed is one year and six months imprisonment which may be extended upto six years and with fine.[51] If any person who has convicted of said offence is again found guilty of that offence he shall be punishable with minimum imprisonment of two years which may be extended to seven years and with fine.

Certain acts such as failure to furnish information required by the State pollution control board, failure to intimate the occurrence of any accident or unforeseen event resulting in pollution of stream and defying the directions of the pollution control boards are punishable with imprisonment for a term which may extend to three months and with fine which may extend to ten thousand rupees or with both.[52]

The act of omission or commission which amounts to an offence of pollution water is provided under section 24 of the Act. It prohibits disposal of polluting matter in streams, well and sewars or on land in excess of the standards established by the state board. Any violation of this section is punishable with the minimum imprisonment of one year and six months which may be extended upto six years and with fine.[53] However, it is to be noted that an act of causing or permitting to cause pollution is an offence only if it is done knowingly. Polluter can escape the liability if he succeeds in proving the want of knowledge. On the other hand, to prove knowledge adds considerably to the prosecuting burden of the state. This dilutes the efficacy of the Act. Pollution whether caused knowingly or unknowingly is a serious matter having disastrous consequences. Thus, in either situation the offence should be made out under the Act. However, if the pollution is caused knowingly the gravity of offence is to be taken more seriously.

Publication of Offenders Names

Very often, the industrialists are not much worried about the fines imposed for the violations of provision of environmental legislations. Nevertheless they are more deterrent about the unpopularity of their banners and public shame. In this context it is interesting to note that the Water Act provides a significant provision which enables the court to publish the offender's name, address, offence committed and punishment prescribed in news papers.[54] This threat of publicity deters the industrialists than any other penal sanction. The expenses of such publication shall be deemed to be part of the cost attending and conviction and shall be recoverable as a fine.

Offences by Companies and Government Departments

Where an offence has been committed by a Company, every person who at the time of offence was committed was in charge of, and was responsible to the company for conduct of the business of the Company, as well as company, shall be deemed to be guilty of the offence and shall be liable to be proceeded against and punished accordingly. However, no such person is guilty, if he proves want of knowledge or that he exercised due deligence to prevent the commission of offence.[55]

Similarly, if the offence is committed by any Department of Government the Head of the Department shall be deemed to be guilty.[56]

Cognizance of offence

The Water Act vests jurisdiction in a court not inferior to that of a Metropolitan Magistrate or Judicial Magistrate of the First Class to take cognizance of any offence punishable under it.[57] A court can take cognizance only if a complaint is made by :

(a) A Board or any officer authorised in this behalf by it; or

(b) Any person who has given notice of not less than sixty days, in the manner prescribed, of the alleged offence and of his intention to make a complaint to the Board or officer authorised as aforesaid.[58]

This provision enables "any person" to move the court complaining of an offence under the Act. However, he can do so only after giving a notice of not less than sixty days to the pollution control board or competent authority, alleging the offence and his intention to make the complaint. This provision to give sixty days notice is to initiate the authority to act upon such allegations and its failing enables the citizen to move the court in this particular regard. There are two things to be taken note of. One is the time gap and the other is how far the competent authority has been catalysed in pursuance of such allegations. The requirement of sixty days notice subverts the very purpose of providing the citizens with the chance to take legal steps against pollution hazards, since sixty days are quite long in the matters relating to the environment. Further, with a view to avoiding prosecution the competent authority may initiate some measures without being properly acting upon such allegations.

Environment (Protection) Act, 1986

Environment (Protection) Act, 1986 (hereinafter referred to as Environment Act) has been passed to give effect to the decisions taken at the Stockholm Conference in June, 1972 relating to the protection and improvement of environment. Prior to the passing of this enactment there was only piece-meal growth of law providing for the protection of some of the components of the environment viz., Water Act, 1974 and Air Act, 1981. These legislations laid emphasis on the control of specific pollution or specific categories of hazardous substances and failed to view the environment in its totality. To remedy this situation the Environment (Protection) Act is enacted. This Act is intended to provide for a comprehensive framework within which almost all issues relating to environment are widely covered. Environment has been defined under the Act as inclusive of inter- relationship that exists among and in between water, air, land and human beings, other living creatures, plants, micro-organisms and property.[59] The responsibility for the protection and improvement of the environment is vested in the Central Government. The Act enables the Central Government to take all measures, as it deems necessary and expedient, for the purpose of protecting and improving the quality of environment and preventing, controlling and abating environmental pollution.[60] In discharging its powers the Central Government can delegate its powers,[61] create authorities or agencies[62] and appoint officers.[63]

The Act enumerates a variety of subjects which include water quality criteria of the receiving streams, standardisation of effluents and location of industries. Exercising the legislative powers conferred under the Act, the central government laid down rules concerning better use of biotechnology, disclosure of information to the persons likely to be affected by accidents in chemical industries and safeguards against transboundary pollution. In order to encourage the industries to take necessary steps for prevention and control of pollution, Central Government instituted a scheme of awards in the form of cup or shield with a citation which will be retained permanently by the awardee. Further, it also iniated a national scheme of labelling "Environment Friendly Products". Any product which is made, used or disposed of in a way that significantly minimises the harm to the environment could be considered as Environment Friendly product and will be certified as eligible for the "ECOMARK". The objectives of the scheme are :

(i) To provide an incentive for manufacturers and importers to reduce adverse environmental impact of products.

(ii) To reward genuine initiatives by companies to reduce adverse environmental impact of their products.

(iii) To assist consumers to become environmentally responsible in their daily lives by providing information to take account of environmental factors in their purchase decisions.

(iv) To encourage citizens to purchase products which have less harmful environmental impacts.

(v) Ultimately to improve the quality of the environment and to encourage the sustainable management of resources.

The Act empowers the Central Government to issue binding directions to any person, officer or authority.[64] For the purpose of achieving compliance with the provisions of the Act. The power to issue directions include the power to direct –

(a) Closure, Prohibition or regulation of any industry, operation or process or,

(b) The stoppage, or rgulation of supply of electricity, water or any other services.

Undoubtedly, this power to give directions serve as a potent weapon in the arsenal of Central Government in controlling pollution.

Apart from making provisions for actual or apprehended environmental pollution, powers of entry and inspection and to take samples, the Act lays down stringent penalty for contravention of the provisions of the Act or rules made or orders or directions issued under it. Strangely, the Act while providing for stringent penalties for the contraventions, contains a provision which almost negates its effect. It suggests that if a particular act or omission amounts to an offence under the Environment Act as well as, say, under the Water Act the offence will be subjected to the punishment prescribed under the latter

Act, and not under the Environment Act.[65] This provision thus nullifies the deterrent effect of the Environment Act. However having realised this folly Water Act is amended in 1988 in order to enhance the penalties in those Acts.

Factories Act, 1948

The Factories Act, 1948 is aimed at securing, among others, the health, safety and welfare of the workers employed in the factory. It imposes a mandatory obligation on the Occupier of the factory i.e., the person who had ultimate control over the affairs of the factory to make effective arrangements for the treatment and disposal of industrial wastes and effluents.[66]

The Factories Act was thoroughly overhauled to meet the changing situation in the wake of Bhopal Gas Tragedy. The imposition of various types of restrictions and conditions in the industrial licenses issues under Industries (Development and Regulation) Act, 1951 were found inadequate. This amendment seek to control hazardous process by imposing more responsibilities on the Occupier of the factory. He is specifically made responsible for the following matters :

(i) to formulate a safety policy and keep it updated continuously;

(ii) to lay down a detailed policy regarding the health and safety of the factory workers and maintain an up to date record of health or medical report of workers;

(iii) to train the workers in handling hazardous substances and their processing;

(iv) to keep a disaster control plan and keep it widely publicised in the neighbourhood and

(v) to constitute in the factory safety committees[67] in which workers also participate.

Workers of the factory are given a right to obtain information relating to health and safety measures at work from the occupier.[68] They can complain about the inadequate or non-existent safety measures to the Factory Inspectorate directly. This provision enables workers to voice against environmental hazards without fear of any ill-motivated disciplinary action of the management.

The Act also provides for the establishment of Site Appraisal Committee[69] which includes the representatives of the Central Board and State Board for the prevention and control of water pollution established under the Water Act 1974. The State Government before granting the permission to start a factory or expand and existing factory involving hazardous substances may take the advisory opinion of the Site Appraisal Committee.

Public Liability Insurance Act, 1991

Increasing use of hazardous substances in a number of industries is fraught with potential dangers to the people at large. Accidents hitherto limited to the workplace and workers employed in the industries can now spill over to the public at large causing injury, sickness or even death. In the past there was no provision for compensation to the public, victims of industrial accidents when they get out of control and affect neighbourhood. The only remedy available to the victims was to initiate legal action in the courts. As we are aware, judicial adjudicatory process is a time consuming and dialatory exercise victims have to wait for a long time to obtain a remedy. This will cause great inconvenience and difficulty to the victims. With a view to providing minimum relief or compensation to the victims of accidents arising from hazardous industries, Indian Parliament recently enacted Public Liability Insurance Act, 1991. It requires the owners of hazardous industries to take mandatory public liability insurance so that atleast minimum scale of relief could be provided to public.[70] The provisions of the Act do not restrict the right of the public to institute legal action, at their choice, to secure further relief.[71] Further it is not necessary that the claimant has to plead and establish that the injury occurred to him was due to the negligence of any person. It is enough if he can show that he has suffered an injury. The scale of compensation or relief to such victims has been indicated in the schedule to the scheme. An application claiming relief may be made by the person who has sustained the injury or whose property is damaged, or by his agent duly authorised. In case of death all or any of the legal representatives of the deceased can make a claim.[72] Applications claiming relief under the Act must be made to the Collector. The Collector after giving the parties an opportunity of being heard and after holding an inquiry into the claim makes an award determining the amount of relief.[73] For this purpose the Collector shall have all the powers of the Civil Court for taking evidence on oath, enforcing the attendance of witnesses, compelling the production of documents and material objects and for such other purposes.

CONCLUSION

It is not a denying fact that industrialisation and its consequent urbanisation create water pollution. But, at the same time, it must also be noted that industrial growth itself is regarded as the yardstick for measuring the progress and civilisation of the country. This complexity calls for a new approach providing for a compromise between the claims of industrial development and environmental protection. It is an imperative to recognise the need for integrating developmental concerns with those about waste disposal and depletion of water resources if the process of development is to be sustained in the long run. The ability to sustain the pace of development depends on preservation of our natural resources such as water, air, soil etc. With these objectives in view the existing legal regimes are to be modified and re-oriented to meet this challenge. A careful study of the relevant provision of various statues including the Water Act 1974, *locus classicus* in the field of water pollution control, reveals that the legislative framework in India is in the process of evolution. The provisions of the old laws which have been casually used for abating water pollution are outdated and fail to provide deterrent effect. The recent legislative enactments such as Environment Act, Public Liability Insurance Act and the recent amendments to the Water Act and Factories Act represent a progressive trend. These laws make an attempt to ensure proper balance between conflicting claims of industrial progress and preservation of our fresh water resources. Pollution control boards are strengthened by conferring upon them extensive powers to issue binding directions. To encourage industries to take appropriate measures to control pollution awards in the forms of cup or shield are introduced. A system of certification of goods that are "Environmentally Friendly" is initiated to make consumers to become environmentally responsible in their daily lives. A minimum scale of relief is assured under the Public Liability Insurance Act to the victims of industrial accidents without dislocating the finances of the industry seriously.

The Draft statement for Abatement of Pollution [74] formulated by the Central Government recently, emphasised the need for reducing the conflict between abatement of pollution and economic development. The modalities recommended by the policy statement include promoting environmentally acceptable production technology, conservation of resources and waste minimisation. Legislation dealing with pollution problems must incorporate and eliminate clashes with environmental criteria. The policy statement also emphasised the need for a procedure for environmental audit in local bodies, statutory authorities and companies to evaluate the effect of their policies on environment.

The Brundtland report[75] which envisaged the futuristic approach of sustainable development highlighted the significance of effective public participation in the environmental decision making process. However, it is rather unfortunate that the existing legislation on water pollution control does not provide for this. There is no compulsion on the authorities to conduct an environmental impact study by affording an opportunity to the general public to participate in that process. The need of the hour is to incorporate the concept of mandatory procedure for environmental impact assessment in the existing pollution control legislations to facilitate effective public participation in the environmental decision making process.

REFERENCES

01. Degradation of the quality of groundwater can be caused by point sources such as septic tanks, garbage disposal sites, cemetries, mine spoils and oil spills. The other possibility of contamination can be by line sources like poor quality water seepage from polluted streams or intrusion of salt water

from oceans. Ecardt C. Beck, "Coping with Groundwater Contamination," U.S.A., Today March, 1984, p. 46.

02. M.A. Acheson, Water Pollution Control in India, C.K. Varshney, Water Pollution and Management (1983) Wiley Eastern Limited, p. 215.

03. The State of India's Environment 1982-A Citizen's Report, Centre for Science and Environment, p. 16

04. Id. at p. 17

05. "The main source of water pollution in India is sewage which comprises 90% of the pollutants, while only 10% come from industry. Many of the cities.... do not have sewage and water treatment plants". Bowonder, *Environmental Management Problems in India*, 10 Environmental Management 599 (1986).

06. Indian Penal Code, 1860 s. 277.

07. *Emperor v. Nana Ram* (1904) 6 Bom. L.R. 52.

08. Section 268 of the Indian Penal Code reads :

"A person is guilty of a public nuisance who does any act or is guilty of an illegal omission which causes any common injury danger or annoyance to public or to the people in general who dwell or occupy property in the vicinity or which must necessarily cause injury obstruction, danger or annoyance to persons who may have occasion to use any public right, A common nuisance is not excused on the ground that it causes some convenience or advantage".

09. Indian Penal Code 1860. s. 425.

10. M.Krishnan Nair, "Legal Control of Water Pollution", [1984] C.U.L.R. 158 at p. 175.

11. *M.C. Mehta* v. *Union of India* AIR 1988 S.C. 1115 The Court observed :

"In common law the Municipal Corporation can be restrained by an injunction in an action brought by a riparian owner who has suffered on account of the pollution of water in a river caused by the Corporation by discharging into the river insufficiently treated sewage."

12. In India, common law doctrine of land ownership is being followed in the sphere of groundwater. Under the common law, any person who owns the (land) surface may dig therein and apply all that is there found to his own purposes. The owner of land accordingly has, as incidental to such ownership, the right to collect store, use and supply to other, the water that is percolating therein. Halsbury's Laws of England, (3rd Ed. 1962). Vol. 34 p. 214.

13. Easements Act, 1882. s. 7 (h)

14. City of Los Angles v. Hunter 156 Cal. 603 at p. 755 (1909) In this case percolating waters are defined as "vagrant, wandering drops moving by gravity in any and every direction along the line of least resistance"; Clinchfield Coal Corp. V. Compton 148 v. 437 (1927). The court defined percolating water as those water which "ooze, seep or filter through the soil beneath the surface without a defined channel or in a course that is unknown and not discoverable from surface indications".

15. Chatfield v. Wilson 28 Vt. 49 (1855) The court observed :

"The Secret changeable and uncontrollable character of underground water and its operations, is so diverse and uncertain that we cannot well subject it to the regulations of law, nor build upon it a system of rules as is done in case of surface streams".

16. Bradford Corp. v. Pickles (1895) A.C. 587. In this case the Court Observed :

"If a land owner sink a tunnel on his land, the effect of which is to cut off the underground water supply from his neighbour, no action will lie even though his object was not to benefit his own land but to extort money from his neighbour by forcing him to buy him out at his own price".

17. In the coastal areas of Saurashtra and Gujarat the ground water is very saline, primarily due to the intrusion of saline sea water following the heavy withdrawal through pumps and wells. A similar situation has developed in Visakhapatnam where the saltwater interface has advanced through structural lineaments and fissures. Again in Coastal Tamilnadu, Over pumping of groundwater resulted in lowering of water table by more than 20 to 30 m during the seventies alone causing the intrusion of oceanic salt water into the aquifer at the rate of 120-150 m/yr. K.S., Valdiya, Environmental Geology : Indian Context (1987) at p. 108.

18. 1990 (1) K.L.T. 580

19. Id. at p. 583

20. Id. at p. 584

21. Ibid

22. Section 91 of Code of Civil Procedure reads : Public nuisances and other wrongful acts affecting the public :

(1) In the case of a public nuisance or other wrongful act affecting, or likely to affect, the public, a suit for declaration and injunction or for such other relief as may be appropriate in the circumstances of the case, may be instituted :

(a) by the Advocate - General or

(b) With the leave of the court by two or more persons even though no special damage has been caused to such persons by reason of such public nuisance or other wrongful act.

(2) ......

23. Halsbury's Laws of England (4th Edn. 1980) Vol. 34. p. 136

24. Supra n. 22

25. Section 133 of Code of Criminal Procedure, 1973 reads :

Whenever a District Magistrate or a Sub-Divisional Magistrate or any other Executive Magistrate, specially empowered in this behalf by the State Government on receiving the report of a police officer or other information and on taking such evidence (if any) as he thinks fit considers (a) that any unlawful obstruction or nuisance should be removed from any public place or from any way, river or channel which is or may be lawfully used by the public :

Such Magistrate may make a conditional order requiring the person causing such obstruction or nuisance or carrying on such trade or occupation or keeping any such goods or merchandise or owning, possessing or controlling such building, tent, structure, substance, tank, well or excavation, or owning or possessing such animal or tree, within a time to be fixed in the order : (i) to remove such obstruction or nuisance; or (ii) to prevent or stop the construction of such substance or if he objects so to do, to appear before himself or some other Executive Magistrate subordinate to him at a time and place to be fixed by the order and show cause in the manner hereinafter provided, why the order should not be made absolute.

27. Id.at p. 1628 per Justice V.R. Krishna Iyer.
28. Id. at p. 1629.
29. Id. at p. 1627.
30. A.I.R. 1979 S.C. 143.
31. 1986 Cri. L.J. 396 (M.P. Indore)
32. 1989 Cri. L.J. 2013 (M.P.)
33. Constitution of India, Article 48 A.
34. Id. Articles 51 A (g)
35. A.I.R. 1985 S.C. 652
36. A.I.R. 1987 A.P. 171
37. Id. at p/. 181
38. 1990 K.L.T. 580
39. Water Act, 1974 s. 2 (j)
40. Id. s. 2 (e)
41. Id. s. 3 and 4
42. P. Leelakrishnan, Statutory Control of Environmental Pollution, [1984] C.U.L.R. 141 at p. 156
43. Water Act 1974 ss. 16 and 17.
44. Id. Section 25
45. Id. Section 24
46. Supra n. 44
47. (1987) 3 Comp. L.J. 69 (S.C.) A writ petition was filed by a public spirited citizen for a writ in the nature of mandamus to the owners of tanneries, restraining them from letting out the trade effluent into river Ganga till such time they put up necessary treatment plants for treating effluents in order to arrest the pollution of the water in river Ganga.
48. Id. at p. 78 per E.S. Venkata Ramaiah, J.
49. Water Act, 1974 s. 32
50. K.V. Ramana Murthy, "Pollution Control : Need for Strategic Enforcement" Vol. 3 Supreme Court Journal (1990) 77 at p. 81.
51. Water Act, 1974 s. 44.
52. Id. s. 41
53. Id. s. 43
54. Id. s. 46
55. Id. s. 47
56. Id. s. 48
57. Id. s. 49
58. Ibid
59. Environment Act, 1986 s. 2(a)
60. Id. s. 3(1)
61. Id. s. 23
62. Id. s. 3(3)
63. Id. s. 4

64. Id. s. 5
65. Id. s. 24 (2)
66. Factories Act, 1948 s. 12
67. Id. Section 41.G
68. Id. s. 41-H
69. Id. s. 41-A
70. Public Liability Insurance Act, 1991 s. 4
71. Id. s. 8
72. Id.s. 6
73. Id. s. 7
74. Drafting Policy Statement for Abatement of Pollution, (1991) Ministry of Environment and Forests, Government of India, New Delhi.
75. World Commission on Environment and Development, Our Common Future (1987).

13

NATURAL DISCHARGE OF GROUNDWATER

V. K. SAXENA

National Geophysical Research Institute,
Hyderabad - 500 007.

INTRODUCTION

There are two types of discharge of water (1) artificial and (2) natural. The artificial discharge is a man made technique in which the continuous discharge of water take place. In natural discharge the groundwater flows through a defined outlet such as spring. Discharge of spring is depends on recharge of water, elevation, geological and hydrological formations. Seasonal changes reflect on discharge rate of springs. Springs water may have good quantity of dissolved solids, varies from hundred to thousands mg/lit. The temperature of the natural discharge is more or less atmospheric temperature but in various cases it reaches even upto boiling (Saxena and Gupta, 1985; Saxena, 1991). Such high temperature natural discharge is known as hot spring or geyser. The thermal springs are very useful for electric power generation, development of various small scale industries (tobacco curing, sugar and alcohol production, aquaculture, pulp processing and catechu purifying etc.), development of health resorts and therapeutic.

NATURE OF NATURAL DISCHARGE

The natural discharge of waters is of three types (1) Cold springs (2) Hot springs, and (3) Geysers.

Cold spring emerges by the gravitational forces in topographic low area preferably in contact zone of two beds of different permeability (Devis and Dewiest, 1966). For cold springs, surface water percolate to shallow depth and emerges on the surface.

GROUND WATER
Editor: P. Sankara Pitchaiah, © 1994 Scientific Publishers, Jodhpur

Hot springs are natural discharge of hot water where surface water temperature is above 5°C of the mean annual temperature. High temperature springs are commonly occurred in places of high geothermal gradients (Saxena, 1987). The aquifers are located usually at a formation which stores the water. The depth of the aquifer may be few meters to kilometers. The water present in the aquifer is heated through some heat source as described latter. Spring water temperature varies from atmospheric temperature to boiling. Steam-water mixture is also present in few places (Saxena and D' Amore, 1984). About 350 natural thermal water discharges are located in various parts of India. This include 175 in Himalaya and remaining in the Konkan coast, Godavari valley and Narmada Sone lineaments, etc. In appearance, the water of hot springs are mainly of two types (a) water clear and transparent, and (b) turbid (Devis and Dewiest, 1966, Hillel, 1971).

Mostly hot spring waters are clear and transparent, but exceptionally turbid in few places of Godavari valley. Turbid water is usually discharge by mud pots. The mud pot is a mixture of water and clay. The undissolved clay particles transport to the surface and form a muddy suspension with the addition of water. Mud pots has less discharge of water.

Geyser is an intermittent water discharge or fountain or columns of hot water, thrown out into surface with force at interval. Geyser rises from few feets to hundred of feets. The presence of mild geyser's activity is noticed in the Manikaran geothermal area of Himalaya. A number of geysers are located in USA and various other parts of the world.

CLASSIFICATION OF SPRINGS

Springs have been classified as depression, contact, seepage, artesian, stratum, tabular or fracture and intermittent etc. (Bowen, 1986).

The depression spring, emerges when the ground surface intersect the water table. The contact spring is a permeable water bearing formation, overlying a less permeable rock, intersects the ground surface. Water emits from many small openings have been termed seepage springs. Such springs discharge little water. The discharge rate of artesian springs are always high and useful for groundwater movement. In artesian springs deep water is released under hydrostatic pressure from confined aquifer. In general the favourable zone to form a spring is a contact zone, between two beds of different permeabilities. The springs are formed in contact zone is known as stratum spring. Spring moves through cannels or fractures in impermeable rock is known as tabular or fracture spring. The tabular spring emerges from water channels or impermeable rock connecting with groundwater. Deep water impeded by underlying impermeable layer in a permeable horizon, emits a spring termed as contact spring. Over flow spring is a permeable bed dip beneath an impermeable zone. The perennial spring derived from extensive permeable aquifers and such spring discharge water in a constant manner. The 'intermittent spring' discharges water periodically and the flow varies from time to time.

ORIGIN AND MECHANISM

Few meters of the top earth crust is affected by seasonal temperature variations and below this, temperature increases with an average rate of 30° C/km. Geothermal gradient is main process of heat transfer from earth's interior to the surface. In areas of old and active metamorphism, the waters are expelled during the recrystallisation and dehydration of hydrous rock minerals under elevated temperature and pressure and encounter as natural thermal water discharge. The favorable conditions for the thermal water discharge are :

(1) Permeable zone/s associated with faults/fissures/cracks (allow water to move to deep levels), and impermeable zone/s of porous materials as to store the water.

(2) Large exposed rock surface area, free movement of water and sufficient time for the heat exchange.

(3) Return path of water to move toward surface favourably through fault/fissure/cracks and permeable strata.

There are various source to supply heat to water. The possible heat sources are shown in Fig. 1. The mechanism and heating behavior of natural discharges are shown by a model as presented in Fig. 2. This model shows water percolate down ward through two different permeable layers along the cracks/fissures. This model storages the water above the impermeable zone, and an aquifer is formed. The water present in the aquifer, absorbs sufficient heat from heat source and with this hydrostatic pressure develope along the route ABCD. The movement path is shown as DE (permeable zone). The path DE is consistent with water/steam. The discharge of water is depend upon relative permeability of CD (up flow path), and EF (out flow route). The out flow path is associated with rock permeability and the pressure of

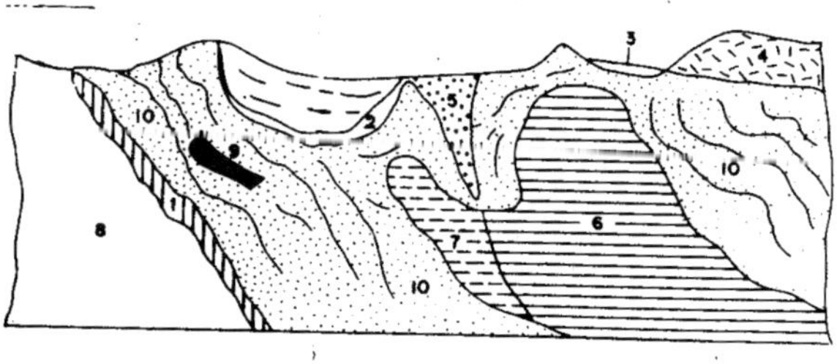

Fig. 1. Possible geological structures responsible to the natural thermal water discharge : (1) Area where the Mantel in thin (2) Geopressured systems (3) High heat flow density under permafrost (4) High heat flow density under an ice cap (5) Volcanic zone (6) Magmatic bodies with residual heat (7) Crypto Volcanoes (8) Deep water circulation (9) Sulphide ore bodies (10) Overthrust and/or subduction zones with enhanced permeability.

the zone DE. This model show a thermo-artesian type system. The intermediate zone DE is controlled by the hot and cold waters density contrast. Mass out flow through EF is normally less than the rate of inflow through CD. In case mass out flow through EF is exceeds, a steam phase develope in DE and the pressure of the steam decreases until the mass out flow through EF is reduced and equal to the mass in flow.

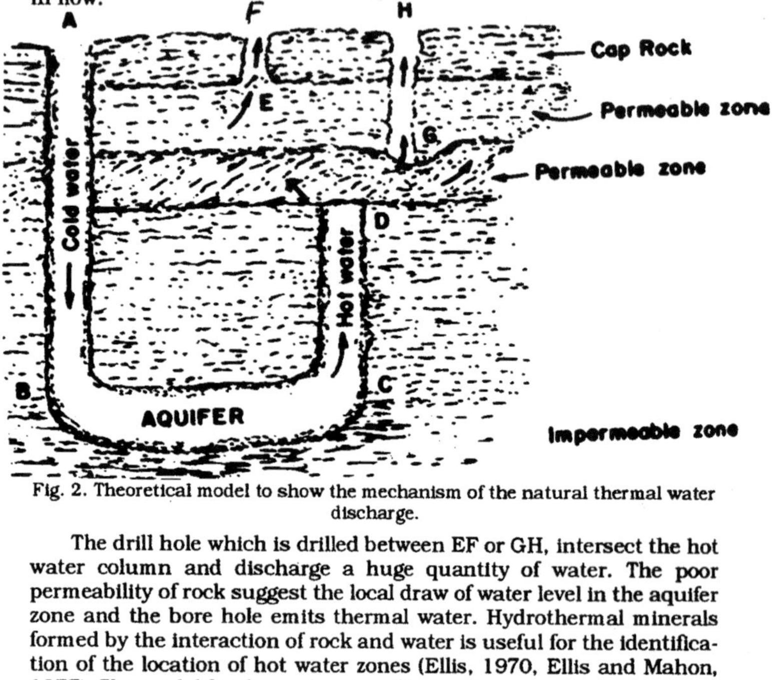

Fig. 2. Theoretical model to show the mechanism of the natural thermal water discharge.

The drill hole which is drilled between EF or GH, intersect the hot water column and discharge a huge quantity of water. The poor permeability of rock suggest the local draw of water level in the aquifer zone and the bore hole emits thermal water. Hydrothermal minerals formed by the interaction of rock and water is useful for the identification of the location of hot water zones (Ellis, 1970, Ellis and Mahon, 1977). The model for the evaluation of hot water zones is shown in Fig. 3. In this model geological formation is crossed by a fault, which transform the pre-existing rock forming minerals and deposite new phases. For this study six drill holes (D1, D2, D3, D4, D5 and D6) of 50-200m depths have been considered. The figure 3 shows schematic of three mineral assemblages i.e. A, B and C which is derived from the primary minerals and the volcanic silica. The C is stable assemblage at high temperature, followed by B and C in order of decreasing temperatures. The D1 and D2 are good producers of hot fluids, whereas D4 and D5 have low enthalpy fluids, and D6 is an unproductive well. The D3 crosses the rocks of high temperature but not intersect the feeder of the hot fluid and thus become unproductive. In this model possibility of the hot fluid is more within D1 and D2.

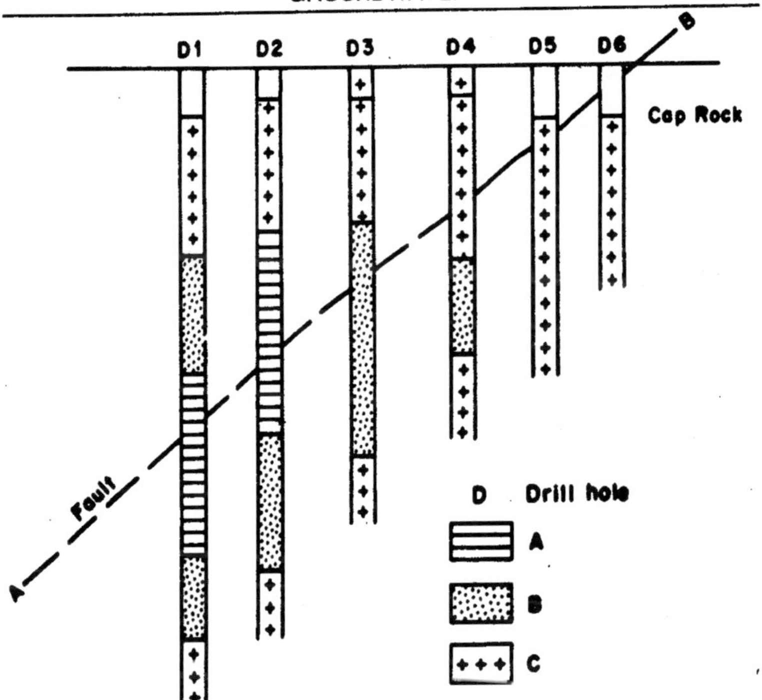

Fig. 3. Theoretical model to show the productive hot water zones.

CASE STUDY

Natural thermal waters of Himalaya, India are shown in Figure 4. The latitude, longitude, surface water temperatures and main rock types around the thermal springs are presented in Table 1. The following six valleys are almost cover the Himalaya's thermal springs.

1. Indus valley : The prominent hot springs of India namely Puga and Chumatang are located in Indus valley, Ladakh at an elevation of 4400 m above mean sea level. The surface temperatures of the spring which are 25 in numbers, varies from 30-93°C. The rock formations near the spring sites are mainly Precambrian paragneisses, schists and phyllites, interlayered with bands of limestone, intruded by granites, amphibolites, serpentinites and tourmaline quartz veins (Fig. 5). The valley floor is covered with alluvium, sandy soil and borax. The sulphur and borax are abundant in the spring site.

2. Parbati valley : Parbati valley is located at an elevation of 2400 m above mean sea level. A large number of hot springs are emerging in this valley. The Manikaran and Kasol are the main hot springs. At Manikaran the temperature of the

Table 1. Locations, surface water temperatures and main features of the natural hot water discharges of Himalaya, India.

Area	Lat. (N) Long. (E)	Tem.	Main rock types	No. of hot water discharges
Puga	33°, 13' 78°, 19'	45–84	paragenesis, quartz, mica schist etc.	50
Chumatang	33°, 22' 78°, 21'	40–82	granite gneisses Indus Flysch	15
Manikaran	32°, 02' 77°, 25'	45–93	quartz, phyllities, granites and mica schist	35
Kasol	32°, 01' 77°, 20'	52–76	quartzites	10
Jan	31°, 59' 77°, 00'	32–34	phyllite, quartzite mica schist	4
Kalath	32°, 12' 77°, 11'	42–46	granite gneisses, and quartz	30
Khirganga	31°, 59' 77°, 31'	50–54	granite gneisses	4
Vashistha	32°, 16' 77°, 13'	57	granite gneisses	2
Suni Tata pani	31°, 14' 77°, 11'	54–56	lime stone, dolomite	4
Tapoban	30°, 29' 79°, 38'	54–61	granite gneisses	5

Table 1 Contd.......

Area	Lat. (N) Long. (E)	Tem.	Main rock types	No. of hot water discharges
Badrinath	30°, 45' 79°, 30'	52–59	granite gneisses	4
Aungi	30°, 45' 78°, 33'	32–36	sand stone	3
Bhukhi	30°, 51' 78°, 41'	44–48	granite gneisses	2
Gangnani	30°, 54' 78°, 41'	61–63	sericite, biotite, schist	2
Joti	30°, 56' 78°, 41'	30–32	porphyritic gneisses	3
Kotimanep	30°, 53' 78°, 14'	34–36	Massive quartzites	2
Jankichatti	30°, 58' 78°, 27'	30–32	sericite, biotite, schist, gneisses	2
Jeori	31°, 36' 78°, 26'	46–48	granite gneisses and quartz	2
Nathpa	31°, 35' 78°, 02'	58–60	granite gneisses and quartz	2
Tatapani Goal	33°, 15' 75°, 07'	38–40	granite gneisses and quartz	2

Temperature in °C

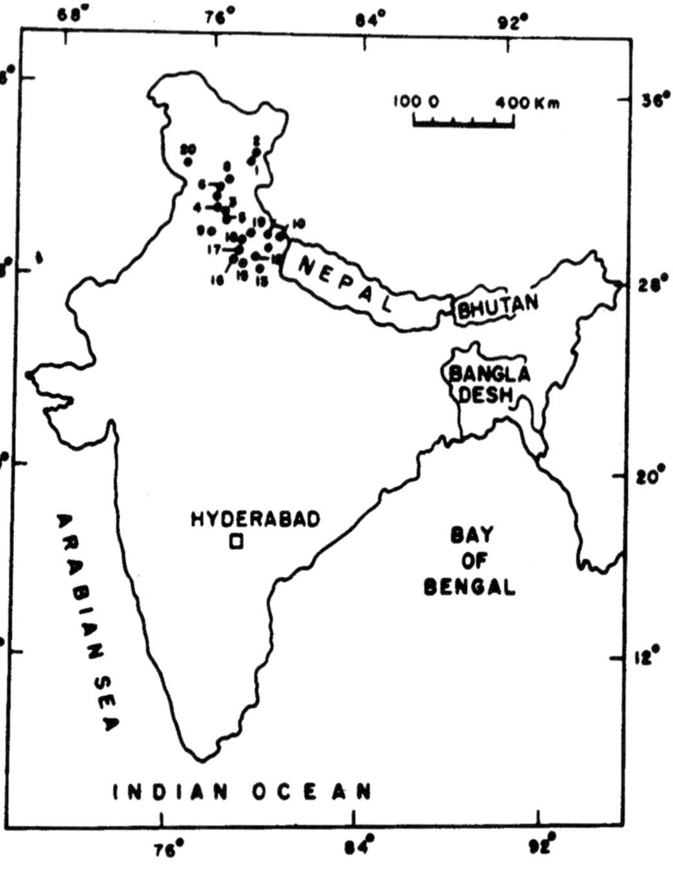

Fig. 4. Location of the natural thermal water discharges of Himalaya region, India.

spring water is 93°C. The rocks are massive quartzites and phyllites. The main geological features are shown in Fig 6. The lithology of Parbati valley is consist with travertine deposits, quartz, granite gneisses and debris. (Saxena, 1993).

3. Beas valley : The surface water temperature of Vashistha thermal spring is 57°C. The rock types of this valley are granite gneisses and quartz.

4. Satlej valley : Various springs (36-60°C), emerges through the interbeded quartzite, gneisses and schists intruded by granites and pegmatites.

5. Alaknanda valley : The well known Badrinath thermal spring (59°C) is present in Alaknanda valley.

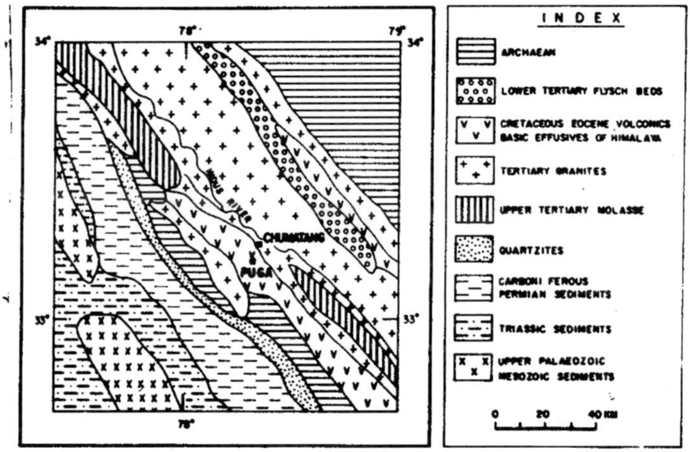

Fig. 5. Geological features of the natural thermal water discharges of the Indus valley.

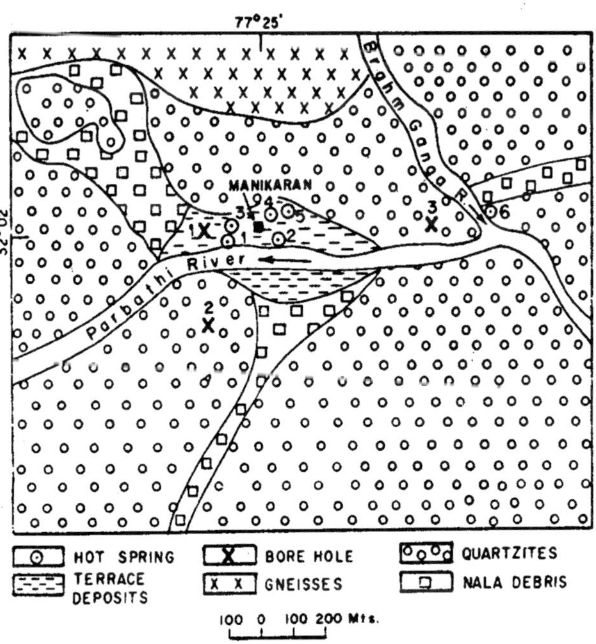

Fig 6. Geological features of the natural thermal water discharges of the parbati valley

6. Dhauli Ganga valley : Tapoban springs (54-61°C) is located in Dhauli Ganga. This valley is mostly covered with granite gneisses and quartz.

The mechanism of Himalayan hot springs and their interaction with rock and the evaluation of aquifer temperature are described as : In hydrothermal system water come in contact with rocks/minerals and carry a large number of ionic species and also few gases. Thus fluid phase moves by convective circulation along with inter-connected faults/pores. This system is chiefly controlled by pressure and temperature gradients. In such process, the hydrostatic pressure impedes the existence of communicating pores or fractures through which the fluid moves. For the aquifer modeling the water rock system is divided into few sub-systems. Each are in equilibrium conditions. In this process temperature, and pressure and chemical potential are kept constant.

The derivation has been made as a example for the aquifer modeling of the potassium feldspar-water :

$$KAlSi_3O_8 + 4H^+ = K^+ + Al^{+3} + 3\ SiO_2 + 2H_2O$$

$$K = \frac{aK^+\ a^{3+}_{Al}\ a35io_2\ a^2H_2O}{aKaAl\ Si_3O_8\ a4H}$$

Where K is the equilibrium constant, and aK^+, aAl^{+3}, $aSiO_2$, aH_2O, aH^+ are the activity of the various phases involved in the equilibrium reaction.

The above equation can also be written as

$$\log k = \log \frac{ak^+}{aH^+}\ \frac{\log a\ Al^{3+}}{aH^{3+}} + \frac{3 \log aSiO_2}{+ 2 \log}$$

$$aH_2O - \log aKAlSi_3O_8$$

Based on equilibrium equations the activity model has been prepared. Activity model is useful, for the study of stability fields for minerals, and equilibrium conditions of the mineral and water. The above equilibrium equation is processed in terms of enthalpy.

$$\log K(T) = \frac{T\Delta Sr^{o}(T) - \Delta Hr^{o}(T)}{2.303 RT}$$

where ΔSr^{o} : standard enthalpy

ΔHr^{o} : enthalpy of reaction

T : temperature in Kelvin

R : gas constant

The change of standard enthalpy and enthalpy at constant pressure are :

$$\Delta Sr^{o}(T) = \Sigma_j S_{jo(T)} - \Sigma_l S^{o}(T)$$

$$\Delta Hr^{o}(T) = \Sigma_j H_{jo(T)} - \Sigma_l H^{o}(T)$$

where j : product

l : reactant

Standard thermodynamical parameters and the heat capacity at a constant pressure ($C^{o}p$). .

$$\Delta H_{o}j(T) = H^{o}Fj(Tl) + \int_{Tr}^{T} CP^{o}J(T)\, dt$$

$$\Delta S_{o}j(T) = \Delta Sj^{o}(T2) + \int_{Tr}^{T} CP^{o}{}_{3} d \log T$$

The heat capacity ($C^{o}p$) is related to temperature. The Maier- Kelley equation is represented by :

$$C^{o}p = a + bt + CT^{-2}$$

where a, b and c are experimental constants.

This equation is used to derive the log K of hydrolysis reaction at different temperatures. The Log K of the dissociation reactions can also be derived by same technique. The above mentioned thermodynamical equations can be solved by computer programme (Kharaka and Barnes, 1973). For the use of aquifer modeling, the ionic concentrations are shown in table 2. The equilibrium constant are referred from Helgesson (1969) and Helgesson et al., (1976). Based on these, activity-activity models are prepared for natural water systems. The activity-activity models for log aK^{+}/aH^{+} vs log aH_4SiO_4 have been drawn at different assumed temperatures. The figures 7 and 8 shows a good equilibrium conditions at 250°C and 150°C particularly for Puga and Chumatang respectively at a constant value of $\frac{a_{Mg}^{2+}}{aH^{2+}}$ (and average values of 10 was used for Puga and 11.1 for Chumatang). The log aNa^{+}/H^{+}/ vs log aH_4SiO_4 activity diagrams have also been drawn (fig. 9 & 10). For Puga, the Puga water apparently equilibrate with adularia, albite, K-feldspar, Na montmorillonite and illite at 250°C and Chumatang at 150°C.

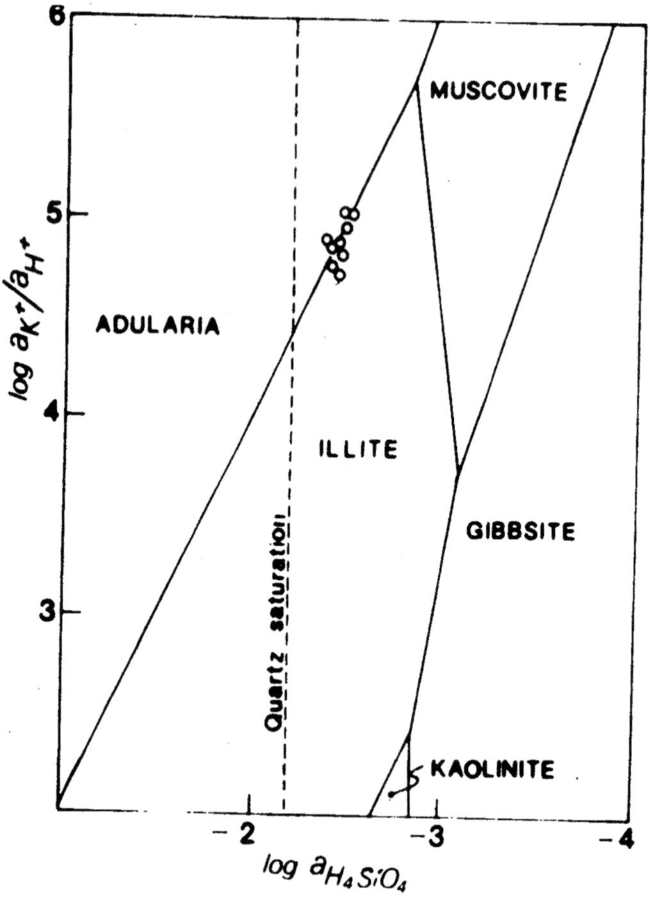

Fig. 7. Activity -activity model for the MgO–K_2O–Al_2O_3–SiO_2–H_2O system at 250°C

Such mineral phases are consistent with the water circulation in metamorphic rocks containing quartz, adularia, illite and clay layers containing montmorillonite and illite. The Na-K activity diagram drawn at 250°C (Fig. 11) also show an equilibration with albite, Na-feldspar at 250°C. The similar models have also prepared for Parbati valley. The log aNa^+/aH^+ and log aH_4SiO_4 show, Manikaran waters are in equilibrium with kaolinite, montmorillonite and quartz (Fig 12) at 150°C. Similar study was carriedout for the remaining natural thermal waters

Table 2. pH, TDS, SiO_2, Na, K, Ca and Mg of natural thermal water discharges of Himalaya, India.

S. No.	Area	pH	TDS	SiO2	Na	K	Ca	Mg
1	Puga	7.8	2090	185	530	65	15	3
2	Chumatang	7.6	1254	166	260	31	18	2
3	Manikaran	7.4	510	120	88	18	32	10
4	Kasol	7.5	414	65	50	11	50	14
5	Jan	7.5	961	32	180	21	68	57
6	Kalath	7.4	1162	38	480	35	38	8
7	Khirganga	7.2	1270	60	350	15	29	7
8	Vashistha	7.4	757	92	200	14	4	17
9	Suni Tata pani	7.5	606	48	205	10	146	30
10	Tapoban	7.2	340	50	19	5	32	16
11	Badrinath	7.3	1840	90	423	35	85	15
12	Aungi	7.2	910	30	170	28	59	61
13	Bhukhi	7.2	992	90	290	16	37	5
14	Gangnani	7.3	807	30	210	29	62	8
15	Joti	7.4	765	60	130	26	59	21
16	Kotimanep	7.2	780	20	180	30	37	45
17	Jankichatti	7.3	440	30	108	-	30	23
18	Jeori	7.6	2120	85	940	36	184	4
19	Nathpa	7.5	750	103	228	14	20	8
20	Tatapani Goal	7.4	956	50	240	8	68	7

chemical compositions are in mg/l

TDS : Total dissolved solids   pH : log H^+ at 25° C

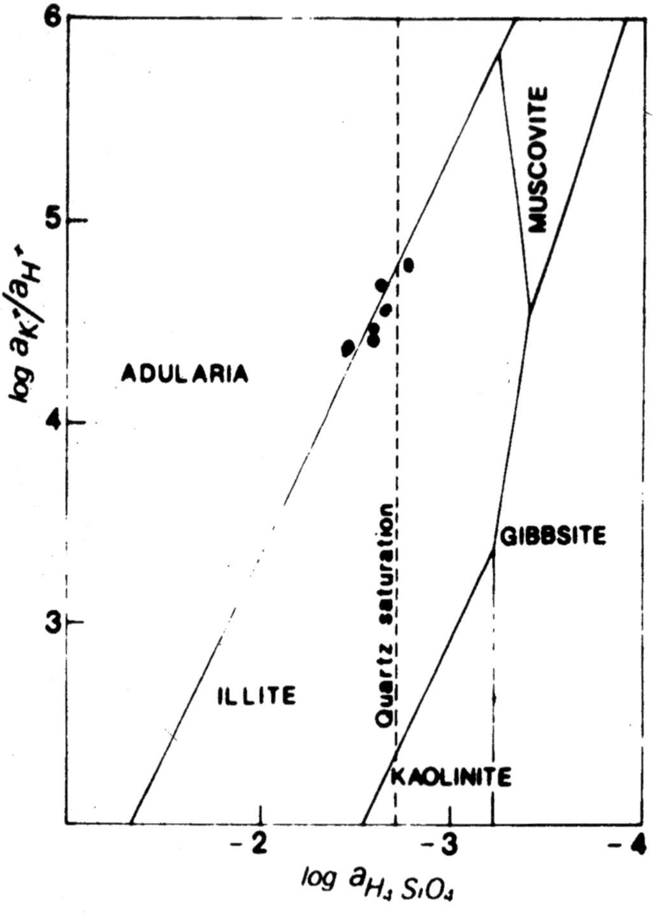

Fig 8. Activity -activity model for the $MgO-K_2O-Al_2O_3-SiO_2-H_2O$ system at 150°C

of the Himalaya. Activity models (Fig. 13 & 14) show that most of the remaining thermal waters of Himalaya are apparently in equilibrium or within the stability field of quartz and clay minerals at 100°C. Chemical indicators for evaluating the aquifer tempeature is very popular during lost two decades. The main geochemical thermometers are silica and NaKCa.

SILICA THERMOMETER : The silica thermometer is based on the temperature dependence of the solubility of silica minerals in water.

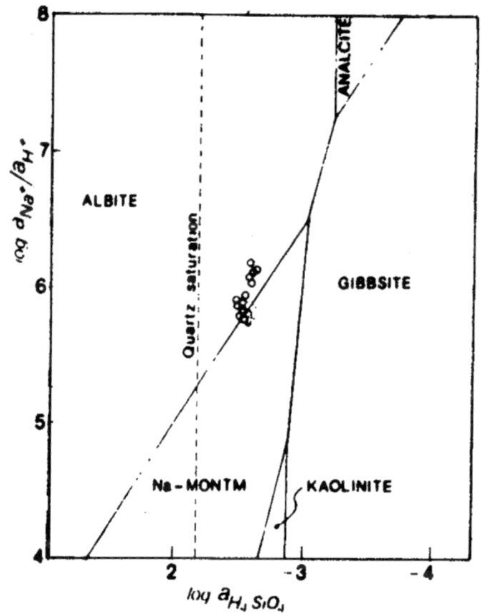

Fig. 9. Activity -activity model for the MgO–Na2O–Al2O3–SiO2–H2O system at 250°C

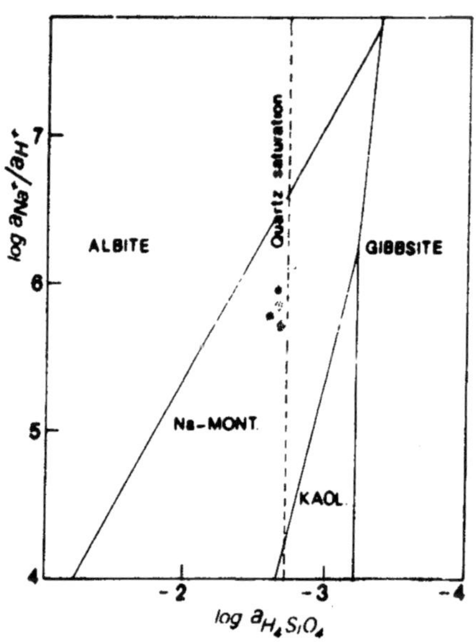

Fig 10. Activity -activity model for the MgO–Na2O–Al2O3–SiO2–H2O system at 150°C

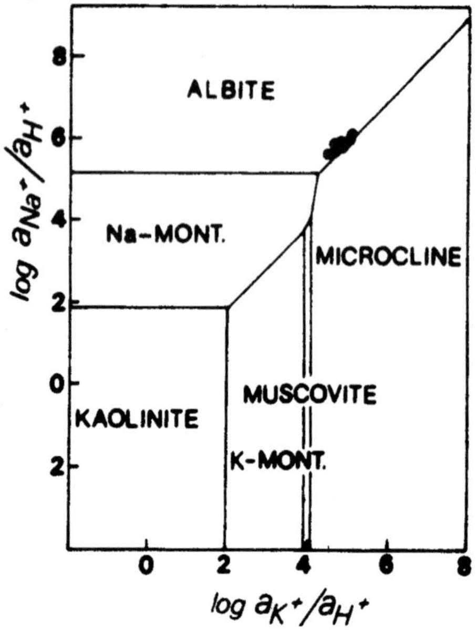

Fig 11. Activity -activity model for the Na_2O–K_2O–Al_2O_3–SiO_2–H_2O system at 250°C

Silica occurs in water in various forms i.e. quartz, chalcedony, cristobolite and amorphous silica. This thermometer was first used by Bodvarsson (1960) for the estimation of subsurface temperature in Iceland. For silica thermometer, a very useful calibration curve was prepared by Fournier and Truesdell (1970). Subsequently Truesdell (1976) expressed it through the following three equations :

$$T_{SiO_2} = \frac{1315}{5.205 - \log S} - 273.15 \quad \text{.... (1)}$$

$$T_{SiO_2} = \frac{1533.5}{5.768 - \log S} - 273.15 \quad \text{.... (2)}$$

$$T_{SiO_2} = \frac{1015.1}{4.655 - \log S} - 273.15 \quad \text{.... (3)}$$

Where T_{SiO_2} is the reservoir temperature in °C.

S= dissolved silica in mg/l

(1) quartz conductive cooling

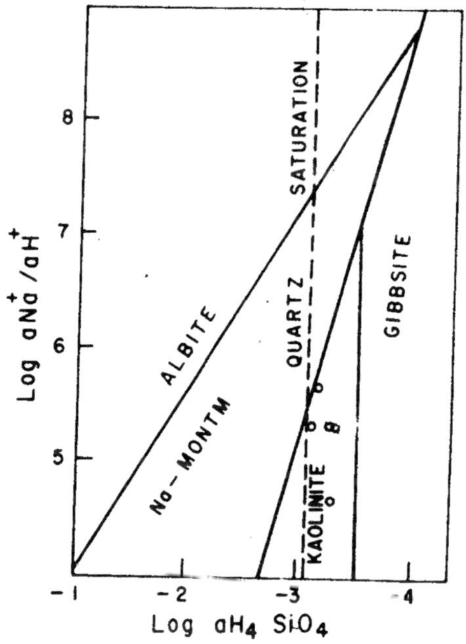

Fig 12. Activity-activity model for the $MgO-Na_2O-Al_2O_3-SiO_2-H_2O$ system at 150°C

(2) adiabatic iso-enthalphic cooling

(3) chalcedony conductive cooling.

Na-K-Ca THERMOMETER : The solubilities of Na^+ 'K^+ and Ca^{2+} are the temperature dependent. The high temperature waters usually consist with more K^+; less Na^+ and Ca^{2+} compared to cold or warm waters. On the basis of bore hole's, bottom temperature and water chemical analysis, the calibration curve has been prepared and a emperical relationship has been derived (Fournier and Truesdell, 1973). This was very much accepted for evaluating the reservoir temperature of the thermal water systems in almost all part of the world. This thermometry has various limitations particularly, Na^+ and K^+ may suggest high temperature for waters high in Ca^{2+} and does not work at temperature around 100-120°C (Fournier and Truesdell, 1973).

Considering the influence of Ca^{2+} in alumino silicate reaction, Fournier and Truesdell (1973) proposed the following emperical NaKCa thermometer :

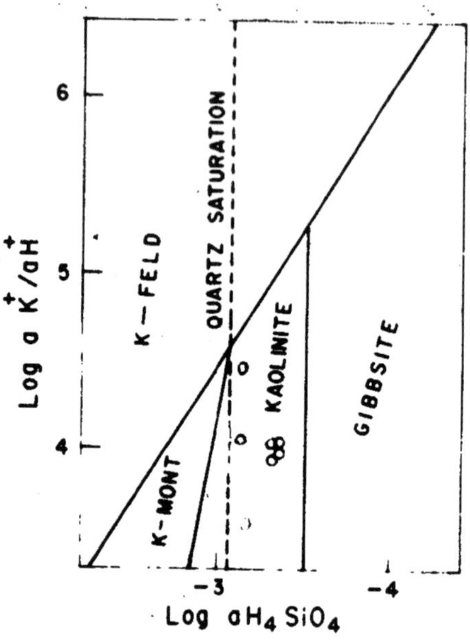

Fig. 13. Activity -activity model for the $MgO-K_2O-Al_2O_3-SiO_2-H_2O$ system at 100°C

$$\log Na/K + \beta \log\sqrt{\frac{Ca}{Na}} = 1647/T^{o}C - 2.24 \quad (5)$$

or

$$T^{o}C = \frac{1647}{\log(Na/K) + \beta \log\sqrt{\frac{Ca}{Na}} + 2.24} - 273.15 \quad (6)$$

The value of β is chosen depend upon wheather the equilibrium of water occurs above or below 100°C and

$$\beta = 4/3 \text{ for } Ca/Na > 1 \text{ and } T < 100^{o}C \quad (7)$$

$$\beta = 1/3 \text{ for } Ca/Na < 1 \text{ or } T \text{ for } \beta = 1/3 > 100^{o}C \quad (8)$$

where Na, K and Ca concentrations are in Mg/l

According to Fournier and Trucsdell (1973), it is observed that these thermometers estimate 15 to 20°C ± of the actual aquifer temperature. The aquifer temperatures of natural water discharge of Himalaya have been evaluated by chalcedony, quartz and NaKCa thermometers and the results are reported in Table 3.

Table 3. Aquifer temperatures of the natural thermal water discharges of the Himalaya, India by T NaKCa and T SiO_2 geothermometers.

Area	T NaKCa	SiO_2 Quartz	SiO_2 Chalcedony
Puga	218	174	-
Chumatang	200	167	-
Manikaran	-	148	-
Kasol	75	-	84
Jan	-	-	50
Kalath	-	-	58
Khirganga	-	-	80
Vashistha	-	133	-
Suni tatapani	-	-	68
Tapoban	-	-	70
Badrinath	-	101	-
Aungi	-	-	46
Bhukhi	154	135	-
Gangnani	-	-	48
Joti	-	-	80
Kati manep	-	-	40
Janaki Chatti	-	-	48
Jeori	144	132	-
Nathpa	159	139	-
Tatapani Goal	71	-	70

Temperature in °C – Not applicable

From this table more than 200°C reservoir temperature was found at Puga and Chumatang, about 150°C at Manikaran, Bhukhi, Jeori, Natpha and about 130°C at Vashistha, Badrinath. In remaining places the aquifer temperature was found less than 85°C.

ACKNOWLEDGMENTS

Thanks are due to Prof. Mario, Fanelli (International School of Geothermics, Pisa, Italy) and Dr. J.G. Negi (NGRI) and Dr. V.Divakara Rao (NGRI) for their support to complete this work. The author is

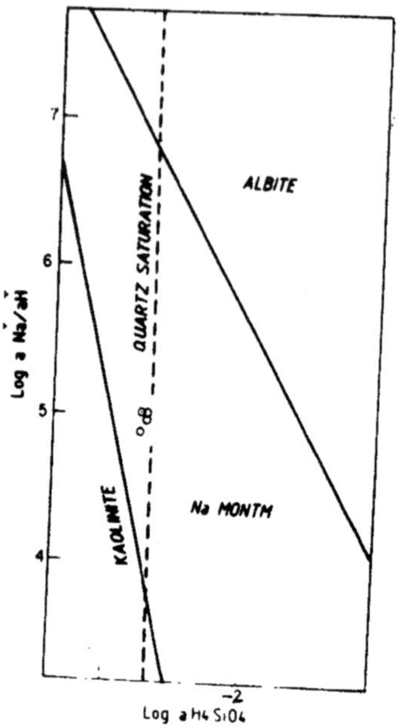

Fig 14. Activity-activity model for the $MgO–Na_2O–Al_2O_3–SiO_2–H_2O$ system at 100°C

grateful to Director, National Geophysical Research Institute, hyderabad, for his kind permission to publish this paper.

REFERENCES

Bodavarsson, G. 1960 Exploration and Exploitation of natural heat in Iceland. *Bull.Volc*, 23, pp 119-128.

Bowen, R. 1986 *Ground water*, 2nd edi., Elsevier Applied Science Publ. Newyork, pp 73-74.

Devis, S.N. and Dewiest, R.J.M 1966 *Hydrogeology*, John Wiley and Sons. Inc, Newyork, pp 463-465.

Ellis, A.J. 1970 Quantitative interpretation of chemical characteristics of hydrothermal system, *Geothermics*, 2, pp 516.

Ellis, A. J. and Mahon, A.J. 1977 *Chemistry of geothermal system*, Academic press, Newyork.

Fournier, R.O. and Truesdell, T.A.H. 1970 Chemical indicator of surface temperature, application to hot spring of yellostone National Park, USA, *Geothermics*, 2, pp 529.

Fournier, R.O. and Truesdell, T.A.H. 1973 *An emperical NaKCa geothermometer for natural waters, Geochim-Cosmocim Acta*, 37, pp 1255.

Helgeson, H.C. 1969 Thermodynamics of hydrothermal system at elevated temperature and pressure . *Am. J. Sci*, 263, pp 729-808.

Helgeson, H.C. Delany, J.M. Nesbitt, H.W. and Bird, D.K. 1976 Summary and critique for the thermodynamical property of rock forming minerals, *Am J. Sci*, 278, pp 1-229.

Hillel, D. 1971 *Soil and water*, Academic press, Newyork, pp 228.

Kharaka, Y. K. and Barnes, J. 1973 *Solmeneq solution mineral equilibria*, US, Geol. Surv. Computer contribution.

Kashef, A.A.L. 1987 *Ground water engineering*, Mcgrw Hill Book Company, Newyork, pp 40-47.

Saxena, V.K. and D'Amore, F. 1984 Aquifer Chemistry of the Puga and Chumatang high temperature geothermal system in India. *J.Volc. Geoth. Research*, 21, pp 333-346.

Saxena, V.K. and Gupta, M.L. 1985 Aquifer Chemistry of thermal waters of the Godavari valley, India. *J.Volc.Geoth.Research*, 25, pp 181-191

Saxena, V.K. 1987 Application of water and gas Chemistry to various geothermal systems of India. *J. Geol. Soc. India*, 29, pp 510-517.

Saxena, V.K. 1991 Study of mineral water equilibria in geothermal waters of Godavari valley, India. *Water Ecology, Pollution and Management*, Edited by P.S. Pitchaiah and B.S. Rao, Chug Pub., Volume 1, Chapter 6, pp 109-124.

Saxena V.K. 1993 Geochemical studies of hot springs of Manikaran H.P., India, *Inland water resources*, : India Edited by M.K. Durga Prasad and Sankara Pitchaiah, P. Ajanta Publication, New Delhi. pp 62-69.

Truesdell A. H. 1976 Summary of section 3-Geochemical techniques in exploration. *Proc. of 2nd UN symp. of the Deve and use of geothermal resources*, Sanfrancisco, US Govt. printing office, Washington, pp liii-lxxix.

14

GROUNDWATER POLLUTION

M.N. BABU & Dr. P. SANKARA PITCHAIAH

Department of Geology, Nagarjuna University,
Nagarjuna Nagar-522 510 (A.P.).

INTRODUCTION

Water is a precious natural resource which is inevitable for drinking and other domestic uses, irrigation, industrial applications, hydroelectric power generation, and waste disposal. Being underground, groundwater is less visible than other water resources and is usually of excellent in quality due to the natural filtration that occurs in the layers of soil through which the water slowly moves.

Pollution of groundwater is the natural, physical, and chemical changes due to human activity, so that the water is no longer fit for a use for which it had previously been suited (Todd, 1980; Paliwal, 1991). The terms pollutant and contaminant require definition. Pollutant is a term which is applied usually for non- living, man made substaces or other nuisances and it refers to their being in excess in a particular location. Often the same substance produced by natural causes may sometimes occur in far greater quantities. Contaminant refers to something which causes a deviation from the normal composition of an environment. Contaminants are not classified as pollutants unless they have some detrimental effect. In pollution, pollutants from different sources have a significant role. Trace metals like Cu, Pb, Zn, Hg, Co, Ni, Mn, certain radioactive isotopes, nitrate, phosphorous, sodium and other elements, also pathogenic bacterias and viruses are the major pollutants. The problem of pollution in groundwater is much less than that of surface water, eventhough this problem is nowadays becoming a severe threaten to public health.

GROUND WATER
Editor: P. Sankara Pitchaiah, © 1994 Scientific Publishers, Jodhpur

MECHANISM OF POLLUTION

The mechanism of pollution due to various pollutants is different from surface water and timelag between pollution discharge at land surface and to reach groundwater may take several years or decades. Some of the factors which control groundwater pollution include : reactions in the topsoil and vadose zone, effect of soil moisture deficiency, laminar flow of pollutants, specific gravity and viscosity effects and slow movement of effluents (Kakkar, 1986). The degree and extent of pollution largely depends on the quality and concentration of the pollutants and the porosity, permeability, hydraulic gradient and storage capacity of the aquifer besides the time factor. Pollution of groundwater may also due to the spreading of pollutant on the groundwater surface or by the direct injection of pollutants into the aquifer system (Verma, 1989). The extent of groundwater contamination from pollutant entering recharge and discharge zones is shown in figure 1.

SOURCES AND CAUSES OF POLLUTION

In general, the main cause of groundwater pollution is due to the disposal of waste water in relation to it's use; which are domestic wastes from urban and rural areas, agricultural wastes, industrial wastes and acid mine drainage and the drainage from beneficiary plants. In addition to these, unplanned nuclear waste disposal is an important source of groundwater pollution. The pollutants from different sources are given in the table below (Table 1.)

Table 1. Classification of sources of Groundwater Contamination (after USEPA, 1977).

Wastes		Non-Wastes
Sources designed to discharge waste to the land and groundwater	Sources that may discharge waste to the land and groundwater unintentionally	Sources that may discharge a contaminant (not a waste) to the land and groundwater
Spray irrigation	Surface impoundments	Buried product storage tanks and pipelines
Septic systems, cesspools etc.	Landfills	Accidental fills
Land disposal of sludge	Animal feedlots	Highway deicing salt stockpiles
Infiltration or percolation basins	Acid mine drainage	Ore stockpiles
Waste disposal wells	Mine spoil piles and tailings	Application of highway salt
Brine injection wells		Product storage ponds Agricultural activities

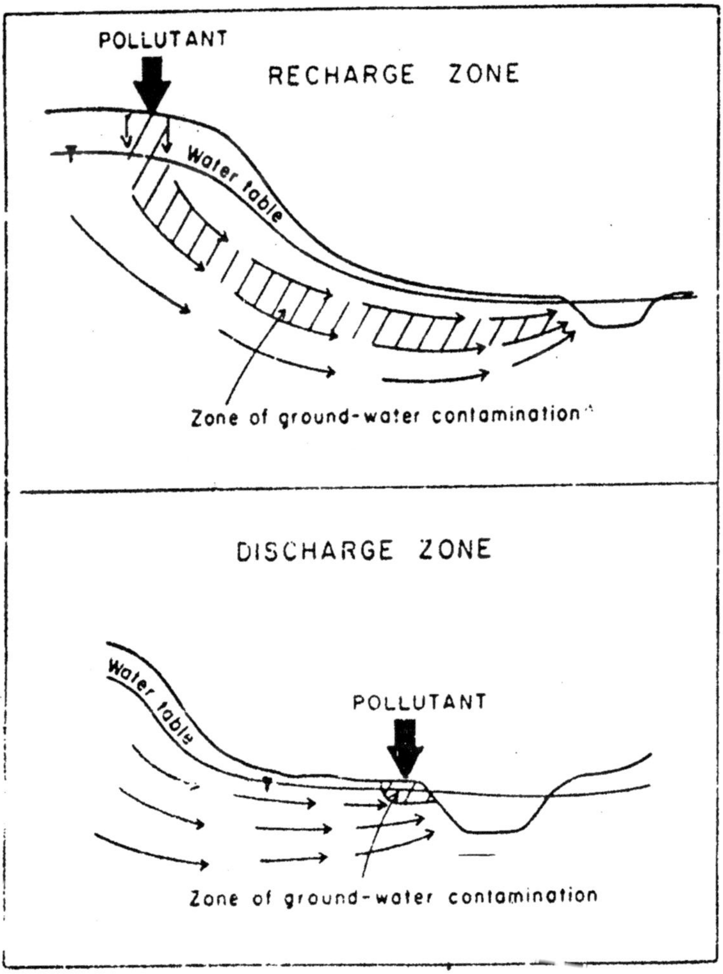

Fig. 1. Extent of groundwater contamination from pollutant entering recharge and discharge zones (after Lindorff and Cartwright, 1977).

The different sources of groundwater pollution are described in the following headings :

(1) Atmospheric dissolved gases

(2) Domestic and Urban wastes,

(3) Agricultural Pollution,

(4) Mining and Industrial Pollution,

(5) Saltwater Intrusion,

(6) Trace metal Pollution, and

(7) Radiological Pollution.

ATMOSPHERIC DISSOLVED GASES

The increase of the concentration of dissolved gases viz, nitrogen, oxygen, hydrogen, carbon dioxide(CO_2), ammonia (NH_3), sulphur dioxide (SO_2), hydrogen sulphide (H_2S) etc. in the atmosphere dissolve in rain water during precipitation will reaches surface water and then to groundwater. The atmospheric precipitation giving very less concentration of mineral matter. Poisonous gases like ethane, methane, ammonia, sulphur dioxide, hydrogen sulphide are mainly coming from the industrial smoke. Absorbed gases in groundwater originated mainly from magma.

DOMESTIC AND URBAN WASTES

About 80% of water pollution in a developing country like India is caused by domestic wastes (Widyanto et. al. 1975). The forms of domestic wastes which adversely affect groundwater quality are sewage and solid wastes. Sewer leakage include high concentrations of BOD, COD, nitrate, organic chemicals and possibly bacteria enter into groundwater (Todd, 1980; Nathanson, 1986). As in the case of many cities, sewage is discharged on agricultural lands without adequate treatment, groundwater may be polluted (Kakkar, 1986).

The disposal of domestic waste water is accomplished in many areas through the use of septic tanks and its leaching field system. In septic tank system it stores the solid waste in a tank and the liquid effluent from the tank flows into a network of buried perforated pipes. The perforated pipe forms a leaching or absorption field which allow the sewer effluent to be absorbed by the soil. This liquid waste in the soil can seep through the vadose zone to the water table. This disposal system leading the source of feacal bacteria and virus contamination. Also the septic tank cleaning fluids contain organic solvents like trichloroethylene (TCE). Other contaminants that can reach the groundwater from septic effluents include detergents, nitrates and chlorides (Nathanson, 1986).

Septic tanks are most likely to contribute to groundwater contamination in areas where (1) the high density of houses with septic tanks, (2) the soil layer over permeable bedsoil is thin, (3) the soil is extremely permeable, such as ground, or (4) the water table is with high population density should not be served with septic tanks and areas with thin soils, extremely permeable soils and high water tables should be avoided (Fetter, 1988).

The land disposal of solid wastes known as landfill creates an important source of groundwater pollution. A landfill may be defined as 'any land area serving as a depository of urban or municipal solid waste (Tods, 1980). Burial in a landfill is the most common means of disposing of municipal refuse, ashes, garbage, leaves demolition debris, and sludges from municipal and industrial wastewater treatment facilities. Radioactive, toxic and hazardous wastes have also been subjected to land burial as a means of disposal. Precipitation that infiltrates the waste can mix with liquid already present in the waste and leach the components from the solid waste. The result is a liquid known as leachate (Fetter, 1988). When leachate from a landfill mixes with groundwater it forms a plume that spreads in the flow of groundwater. The problem of pollution from landfills is greatest where high rainfall and shallow water table occur. The volume of a leachate that is produced is a function of the amount of water percolating through the refuse. The location, nature and magnitude of the pollution depends on local conditions such as size of the landfill, the age of the landfill, and amount of the refuse, the hydraulic gradient and the geologic strata (Olania, 1977).

Important pollutants frequently found in leachate include BOD, COD, iron, manganese, chloride, nitrate, hardness, and trace elements. Hardness, alkalinity and total dissolved solids are often increased, while generation of gases such as methane, carbondioxide, ammonia, and hydrogen sulphide are further biproduct of landfills (Todd, 1980.)

AGRICULTURAL POLLUTION

Groundwater pollution due to agricultural sources is mainly derived from fertilizers, irrigation salts, animal wastes and crop waste disposal (Kakkar, 1986). Approximately one half to two third of the water applied for irrigation of crops is consumed by evapotranspiration; the remainder, termed irrigation return flow, drains surface channels or joins the underlying water.

The degradation results from the addition of salts by dissolution during the irrigation process, from salts added as fertilizers or soil amendments, and from the concentration of salts by evaporation. Animal wastes may transport salt, organic loads, and bacteria into the soil. Storm run-off in contact with the manure carries excess NO_3 to groundwater (Tamta, 1991).

When fertilizers are applied to agricultural land, a portion usually leaches through the soil and to the water table. The primary fertilizers are compounds of nitrogen, phosphorous, and potassium. Excessive use of fertilizers and pesticides for agricultural development may pollute groundwater resources. Many of the pesticides and herbicides are cumulative poisons (Verma, 1989). The impact of peaticides on the groundwater quality depends on the properties of the pesticide residue, rainfall or irrigation rates, and soil characteristics.

MINING AND INDUSTRIAL POLLUTION

Mining Effluents

Mining activities are long and complex process include several steps in between sampling of ores to concentration and excavation of fluid products. Extraction and processing of metallic ore and coal has been the source of both surface and groundwater contamination. Surface water flowing in the vicinity of the mines can pick up dissolved metals and other solids, acidity and even radioactive substances. As it infiltrates the earth, either in the open pits from strip mining, or in underground tunnels and shafts, the polluted water easily carries the contaminants into underlying groundwater aquifers. This will lead to the pollution of groundwater. In opencast and underground mining several tons of waste materials are removing and dumping somewhere near to the mine area. This artificial barren hills of waste dumps exposed to wind and water and consequently produce disperson of metals through natural transporting media. Groundwater moving through mineralised rock zones may contain excessive amounts of heavy metals. Disperson of metals not only depend on mining activities but also on the physico-chemical characteristics of the deposits, E_H-pH conditions of the transporting medium, geochemical mobility of the metal concerned, precipitation barriers enroute, so on and so forth (Sahu, 1990). Leaching contaminants from tailing ponds and slag piles is also a source of groundwater contamination from mining operations.

The precise effect of pollution will vary with the type of waste discharged, but there is an increase of Total Dissolved Solids noticed in the mine drainage waters. pH of the receiving waters is also brought down which leads even to dissolution of some materials in water (Trivedy, 1990). The waste waters have the following characteristics (i) low pH, (ii) generally high content of heavy metals like Fe, Mn, Al etc. (ii) high dissolved solid content (iv) high sulphate content amongst anions, and (v) relative absence of organic matter.

The waste water from coal mines may be broadly divided into two categories : (i) mine drainage waters and (ii) acid mine drainage waters (Handa, 1990).

Mine drainage waters : This is the mining water pumped out during drainage operations. The underground mines that reach below the water table of an unconfined aquifer or interact with a confined aquifer, necessitate pumping out of groundwater seepage into the mine.

Acid Mine Drainage : This is the wastewater produced in coal mining operations which owe their origin to the oxidation of sulphur bearing materials dumped near coal mines.

Mining and milling expose overburden and waste rock to oxidation. Ore beneficiation plants, concentrators, and smelters normally located near mine heads produce a large amount of tailings or slags which are either discharged into nearby aquatic channel, tailing dams or heaped around the cinders. The concentrators invariably use a large amount of water, a considerable proportion of which is diverted in hydraulic transport of the tailings into dams and finally into surface and subsurface waters.

Industrial effluents

Industrial wastes disposed of (the Indian standards for industrial effluents discharge are given in the table 2.) in surface impoundments, such as landfills or lagoons, represent a most significant source of groundwater contamination. Rapid industrialisation without provision of proper treatment and disposal of waste and effluent is another source of pollution. Industries such as sugar factories, textile mills, distilleries, tanneries, mineral and metal processing industries may generally cause pollution. The waste products of industries contain hazardous substances such as arsenic, hexavalent chromium, cadmium and cyanide in toxic concentrations. A large fraction of the hundreds of millions of tons of industrial wastes generated each year is hazardous. Land disposal of these liquid and solid wastes is practiced because it is least expensive way to get rid of the unwanted materials (Nathanson, 1986). Several of the inorganic and organic constituents present in industrial effluents are toxic. The effluents from industries are either discharged into pits or are passed through unlined channels to nearby depressions where several toxic constituents percolate into groundwater.

Industrial wastes are sometimes pumped into the groundwater under pressure through deep wells, in a process called deep well injection. This is generally an acceptable method for industrial waste disposal, if the geological conditions are suitable. But even with deep well injection, accidental contamination of important water supply aquifer is possible.

SALTWATER INTRUSION

Sea water intrution in confined aquifers is a common problem in the coastal areas. Several mechanisms such as the Ghyben-Herzberg and negative gradient mechanisms cause the sea water to enter the aquifers (Chandrasekharam, 1993). Because of the increasing population, urbanisation, and industrialisation, increasing quantities of groundwater are being used. Therefore saltwater may invade freshwater aquifers to create point or diffuse pollution sources. The mixing of seawater in the openwells take place whenever the water table fall below the mean sea level. In coastal aquifers seawater is the pollutant, while in inland aquifers underlain salinewater may be responsible (Todd, 1980).

Table 2. Indian standards for industrial & sewage effluents discharge.
IS : 2490-1982.(All values in mg/1).

S. No.	Parameters	Industrial Effluent		Into marine coastal area	Into Public sewage	Sewage surface
		in to inland surcace water	on-land for irrigation			
1.	Colour & odour	-	-	-	-	-
2.	Suspended solids	100	200	a)for process waste water 100 b) For cooling water effluents 10% above total suspended matter of influent cooling water.	600	30
3.	Particle size suspended solids	850 micron	-	a)Floatable Solids max. 3mm. b)Settleable solids max. 850μ.	-	-
4.	Dissolved solids	2100	2100	-	2100	-
5.	pH	5.5-9	5.5–9	5.5–9	5.5–9	-
6.	Temperature (°C)	45°C	-	-	-	-
7.	Oil & Grease	10	10	20	20	-
8.	Total Residual (Cl^-)	1	-	1	-	-
9.	Nitrogen (as N)	50	-	50	50	-
10.	Kjeldahl Nitrogen	100	-	100	-	-
11.	NH_3	5	-	5	-	-
12.	BOD	30	100	100	350	20
13.	COD	250	-	250	-	-

Table Contd...

14.	As	0.2	0.2	0.2	0.2	–
15.	Hg	0.01	–	0.01	0.01	–
16.	Pb	0.1	–	1.0	1.0	–
17.	Cd	2	–	2	1	–
18.	Cr^{6+}	0.1	–	1	2	–
19.	Cr (Total)	2	–	2	2	–
20.	Cu	3	–	3	3	–
21.	Zn	5	–	15	15	–
22.	Se	0.05	–	0.05	0.05	–
23.	Ni	3	–	5	3	–
24.	B	2	2	–	2	–
25.	Na	–	60	60	–	–
26.	Na_2CO_3 (Residual)	–	50	–	–	–
27.	CN	0.2	0.2	0.2	0.2	–
28.	Cl	1000	600	–	1000	–
29.	F	2	–	15	15	–
30.	Dissolved PO_4 (as P)	5	–	–	–	–
31.	SO_4	1000	1000	–	1000	–
32.	Sulphide (as S)	2	–	5	–	–
33.	Pesticides	Absent	Absent	Absent	Absent	–
34.	C_6H_5OH	1	–	5	5	–
35.	Ratio active meterials					–
	a) α emitters μc/ml max	10^{-7}	10^{-8}	10-7	10^{-7}	–
	b) β emitters μc/ml max	10^{-6}	10-7	10^{-6}	10^{-6}	–

TRACE METAL POLLUTION

Almost all the heavy metals found in water are as traces, known as trace metals or trace elements. The term trace element is rather loosely used in current literature to designate the elements which occur in small concentrations in nature. The growing public concern over the deteriorating quality of the environment had lead to a generalised wage when referring to trace elements. It has been known for several decades that trace quantities of certain elements exerts a positive or negative influence on plant, animal and human life (Forstner, 1979). Trace metals though naturally occurring can be present in some areas in sufficient concentrations and physico-chemical forms that might create pollution problems. Sources of pollution of trace metals like Hg, Pb, Cd, As, Cr, Zn, Mn and Fe are mainly aquatic release from industrial operations, atmospheric releases from fossil fuel burning, domestic sewage discharges and land run-off (Pillai, 1979). Natural or manmade sources of metals at or near the land surface may be rendered less important through removal of metals as water percolates through the ground.

RADIOLOGICAL POLLUTION

As a result of the industrial use of nuclear fission particularly for power generation, waste products are formed which may be discharged in surface may pollute the groundwater resources in future. Operations like mining and milling of radioactive ores, chemical reprocessing, oceanic and atmospheric nuclear explosions and radioactive waste disposal from nuclear power plants may cause the contamination of groundwater due to seepage of polluted surface water into the ground (Kakkar, 1986; Verma, 1989). Radioactive minerals are present on earth, in all parts of the world, contribute radioactivity to water which comes in contact with them, causing sometimes what is called as natural radioactivity. Mining operations of nuclear reactors for industrial and other purposes produces wastes of relatively fine materials similar to clays which are often intercharged into stream as means of disposal (Shrivastava, 1986).

PREVENTIVE MEASURES OF GROUNDWATER POLLUTION

The process of groundwater contamination will continue, but it's impact can be reduced. The role of hydrogeologists in regulatory agencies should be strengthened to provide proper evaluation of potential sources of contamination and to aid in remedial action when groundwater is contaminated. Cooperative efforts to develop strategies will ensure proper handling of future emergencies (Lindorff, 1979). The case studies indicating that once groundwater contaminated, remedial action is time consuming and expensive. Therefore protection of groundwater from contamination is essential. To effectively regulate potential sources of contamination, we must understand the behaviour of contaminants in the surface. Then we can predict the environmental impact.

The importance of understanding the hydrogeological conditions to permit evaluation of potential environmental impact has been stressed in recent research programmes. Knowledge of position of the landfill in the groundwater flow system and the nature of the surrounding earth materials provide a basis for determining the degree of natural renovation offered by the site. To be aware of the groundwater resources and the existing or potential uses of local aquifers is also essential. Where geological factors are inadequate to provide for natural renovation of leachate, the site can be engineered to protect the subsurface environment (Lindorff, 1979). The behaviour of other potential contaminants also to be investigated.

Groundwater quality monitoring plays an important role in groundwater protection and quality conservation (Jaroslav, 1991). Groundwater quality monitoring programme defined as scientifically designed surveillance system of continuing measurements, observations, and evaluations if necessary. Monitoring methods may include not only sampling and analyses of groundwater quality, but also determination of groundwater levels and flow directions, measurements of moisture in the unsaturated zone, geophysical surveys, evaluations of wastes and other materials contributing to subsurface pollution, testing of pipeline and tanks, and aerial surveillance (Todd, 1980). Landuse management applied on the local level by town and cities can be effective in preventing aquifer contamination. Strict enforcement of regulations pertaining to the siting, design can reduce or eliminate the incidence of sewage contamination of private wells. Prudent application of pesticides and fertilizers in agricultural areas can also be effective in this regard (Nathanson, 1986).

The waste water coming out of industries and mining must be treated by various physico-chemical techniques depending on the type of pollutant generated (Trivedy, 1990). The various physico-chemical techniques devised for removal of chemical, biological or radiological pollutants involve adsorption, electrodialysis, ion exchange and reverse osmosis.

Pollutants in groundwater tend to be removed is reduced in concentration with time and with distance travelled. Mechanisms involved include filtration, sorption, chemical process, microbial decomposition and decomposition (Todd, 1980). Filtration can remove particulate forms of iron and manganese as well as precipitate formed by chemical reactions. Clays, metallic oxides and hydroxides and organic matter function as sorptive materials in sorption, which removes chloride, sulphate and nitrate. Precipitation in groundwater can remove the major constituents like calcium, magnesium, and sulphate. Volatilisation and loss as a gas apply particularly to reactions involving nitrate and sulphate. Radioactive decay based on the half life of a radio isotope acts as an attenuation mechanism of radioactive pollutants. Dilution is the most important attenuation mechanism for pollutants after they reach the water table.

Recent laws related to solid and hazardous waste disposal will significantly reduce new contamination. Eventhough there is no single federal law for groundwater protection on a national scale. Many of the laws passed over the years regarding water pollution and waste disposal do relate to groundwater pollution to some extent. But not all types of situations are covered and there remain gaps in the regulations regarding groundwater.

CASE STUDIES OF GROUNDWATER POLLUTION IN INDIA

In India, the impact of pollution is generally due to the haphazard urban development without adequate attention to sewage and waste disposal, rapid industrialisation without proper treatment and disposal of waste products, insanitary dumping of refuse and other solid wastes near aquifers, excessive use of fertilisers and pesticides for agricultural development and poor drainage in agricultural soils. Some important case studies are given below :

Groundwater pollution of twentyfive wells near to the open refuse dumps at Jaipur were studied by Olania and Saxena (1977), during 1969 to 1971 in order to assess the extent of groundwater pollution by leachate around the refuse dumps. The results indicate that the pollution by total dissolved solids (TDS), chloride and iron salts is felt considerably upto a distance of about 450 m. The increase in hardness and COD are also perceptible upto about the same distance.

The study of pollution of groundwater in Visakhapatnam area, Andhra Pradesh, by Subba Rao and Krishna Rao (1990) indicated that the intensity of pollution of groundwater was not so alarming upto 1975 in Visakhapatnam area, but after that it rapidly increased due to sweeping growth of industrial and urban activities and lack of environmental protection. Hydrochemical facies in 1975 are characterised by Na-HCO_3 type and in 1981 by Na-Cl type indicating that the salinity in groundwater increased from 1975 to 1981. Also the concentration of such as total alkalinity (TA), Na+K as Na, HCO_3, Cl and SO_2 are increased during the period from 1975 to 1981.

The investigation of phreatic aquifer water in the basaltic plateau of the Godavari basin Karnataka comprising taluks of Bidar, Bhalki and Aurad in Bidar district; evidently polluted by non point source live stock manurial NO_3 in dugwells water (Tamta, 1991).

Saxena and Vijayamohan (1993) reported the excess concentrations of toxic trace elements in deep boreholes at some locations in Hyderabad, Andhra Pradesh. The investigation showing that the concentrations of trace metals like Pb, Ni, Cr and Cd are high in the borehole waters at Uppal area, surrounded by several industries such as medicinals, chemicals, dyes, oil and fats, acids, potteries, textiles, metals and confectionaries which are discharging a large amount of waste material to sewage water.

The saltwater intrusion studies along the coastal aquifers in Malappuram district, Kerala (Chandrasekharam, 1993) showing that bicarbonate and chloride are more in the groundwater. The high concentration of chloride indicating salinity. Water level data indicate that mixing of seawater in the openwells takes place whenever the watertable fall below the mean sea level. In such a condition, due to the prevailing negative gradient, certain amount of seawater mixes with groundwater.

The groundwater pollution studies around North Railway City Station, Lucknow (Singh, et al. 1991) reveal that the area is affected by high nitrate concentration and the main source of nitrate pollution is septic tanks which are located in the study area.

The quality of groundwater in Coimbatore, the major textile industrial city in South India (Someswar Rao et. al., 1993) reveals that the water is hard and saline which is unfit for domestic use. The reason for the contamination of groundwater is found only due to the seepage of textile and dyeing industrial effluents containing high concentrations of sodium.

The studies conducted by Subba Rao and Venketa Rao (1973) revealed that water supplied from an infiltration gallery constructed in the bed of non perennial river SARDA (Ankappalli town, Andhra Pradesh) exhibited severe taste, odour, colour and filamentaneous grouth which was a result of indiscriminate discharge of untreated sugarcane distillery wastewater into the riverbed, as reported by them.

Rameswar D. Verma, (1989) summarised some case studies of groundwater pollution in India are reported below.

In an investigation, Olania et. al. (1978) reported high MPN count of total coliforms and faecal streptococcus, and high concentrations of chlorides, nitrates, and sulphates and COD in water of wells located at 30 to 60 metres (average) distance from sewage forms at Jaipur.

Sridhar and Pillai (1973) reported the pollution of groundwater at Bangalore. They conducted field and laboratory investigations of water samples collected from wells located in the neighbourhood of (i) the swag channels in the outskirts of Bangalore City, (ii) septic tanks in the campus of the Indian Institute of Science and (iii) the septic tanks of four industrial establishments. Results of investigations revealed that all the wells exhibited sewage pollution.

In a geohydrological investigation in northern and south western Haryana localised rise in nitrate concentration from 43 mg/1 at Singhani to 1920 mg/1 in Narnaul groundwater was observed by Kakkar (1981). In the absence of any geological source of nitrate in the area and non uniform distribution of nitrate along the direction of groundwater movement, nitrate pollution is likely to be caused by sewage and agricultural waste.

From the field studies in Ludhiana in Punjab, Kakkar and Bhatnagar (1981) concluded that groundwater at shallow depths of aquifer near bicycle factories had been polluted by hexavalent chromium and cyanides in concentrations ranging from 3.5 to 12.9 mg/1 and 0.05 mg/1 to 0.98 mg/1 respectively. Other trace elements such has copper, zinc, cadmium, cobalt, molybdenum, strontium, lithium and silver were also detected in groundwater in different concentrations, although below the permissible limits for drinking water.

In another investigation analysis of water samples drawn from the hand pumps installed in Industrial Area of Ludhiana, Punjab revealed that ground water had been contaminated by hexavalent 1.75 mg/1, 0.04 to 0.16 mg/1 and 0.71 to 4.0 mg/1 respectively. (Goyal. et. al. 1981).

Well water quality monitoring was carried out by Tanwar (1981) for four years. He observed continued increase in concentration of salts (such as chloride and sulphates of calcium and magnesium) as a result of increased use of fertilizers (4 to 57 kg. per hectare within 12 years) in Haryana. Author suggested that in case no preventive step is taken the groundwater pollution could become a serious problem.

Quality of groundwater in Warrangal in Andhra Pradesh has deteriorated because of land disposal of untreated waste water from tannery and textile mills which found access to groundwater. Fortunately the pollution had travelled only 5 metres from the disposal site and wells located in sandy loam and rocky formations at 150 metres from disposal site were found unaffected (Naram, 1981).

In Tamilnadu also, the tannery waste is posing problem to groundwater quality. Krisnaswami and Haridas (1981) reported the results of groundwater-quality investigations along a seasonal river Palnar that receives organic/inorganic wastewater from tanneries located at Vaniyambadi and Ambur (close to the origin of the river). A significant rise in total dissolved solids, salinity and hardness in wells located near factories led the authors to conclude that preventive measures must be instituted immediately to control groundwater pollution.

BIBLIOGRAPHY

Chandrasekharam, D. 1993. Saltwater intrusion mechanism along the coastal aquifer, Malappuram district, Kerala. pp 1-10. Inland water resources : India, Ajanta Books International, New Delhi. 344 pp.

Fetter, C.W. 1988. Applied hydrology, Merril Publishing Co. U.S.A. 592 p.p.

Forstner, V. and Wittman, G.T.W. 1979. Metal pollution in the aquatic environment, Springler Verlag. 486 p.p.

Goyal, M.R.; O.P. Abrol and A.K.Vohra and withman G.T.W. 1981 "Pollution of upper aquifer in Punjab" Proceedings of an international symposium held in the Netherlands, March 23-27, 1981. pp. 105-110.

Handa, B.K. 1990. Abatement of water pollution in mining industry. pp227-57,

Impact of Mining Environment. Ashish Publishing Co. 310 pp.

Jaroslav, Vrba.; 1991. Groundwater quality and Monitoring Effective Methods of Hydrogeological System Pollution-Prevention. Environ. Geol. Water. Sci. 17, No. 1, Springler Verlag. pp. 9- 16.

Kakkar Y.P. 1981 Nitrate pollution of Groundwater in southern and south Eastern Haryana, India. Proceedings of an international symposium held in the Netherlands, March 23-27, 1981. pp. 125- 129.

Kakkar Y.P. and N.C. Bhatnagar 1981. Groundwater Pollution due to industrial effluent in Ludhiana, India. Proceedings of an international symposium held in the Netherlands, March 23-27, 1981. pp 265-272.

Kakkar, Y.P. 1986. Groundwater Pollution. Groundwater News, 5, No. 11 & 12 pp 51-56.

Krishnaswamy, R and G. Haridas 1981. "Groundwater pollution by tanneries in Tamil Nadu (India). Proceedings of an International symposium held in the Netherlands. March 23-27. 1981.

Lindorff David E, and Keros Cartwright 1977. Groundwater contamination : problems and remedial actions. Illinois Geological Survey. Environmental Notes 81, 58 pp.

Lindorff, David E. 1979. Groundwater Pollution A Status Report. Groundwater. pp. 9-17.

Manual on Water and Waste water Analysis, NEERI, Nagpur, 1988. 331 pp.

Naram, K.R. 1981. "Groundwater pollution in Warrangal Town", Andhra Pradesh, India, Proceedings of an international symposium held in the Netherlands. March 23-27, 1981 pp 163-168.

Nathanson, Jerry A. 1986. Basic Environmental Technology. John Wiley & Sons. 479 pp.

Olania, M.S. & K.L. Saxena 1977. Groundwater Pollution by open refuse dumps at Jaipur. Ind. Jr. of Environs. Health. 19. No. 3. pp 176-188.

Olania, M.S.; W.G.Nawlakhs; H.C. Sharma and K.L. Saxera (1978) Pollution studies of well waters in sewage farms of Jaipur, Jr. of Env. Health 20, No. 4. pp. 398- 412.

Pillai, K.C. 1991. Heavy metals in aquatic environment pp 74-93. Water Pollution and Management. Wiley Eastern Ltd. 242 pp.

Rao, C.S. 1991. Environmental Pollution Control Engineering. Wiley Eastern Ltd. 535pp.

Someswara Rao N. 1993. Groundwater Quality. A case study of textile industrial city, Coimbatore, India pp. 30-37. Inland water resources; India, Ajanta Books International, 344 pp.

Subba D Rao N, & G. Krishna Rao 1990. Intensity of pollution of groundwater in Visakhapatnam area, Andhra Pradesh, India. Jr. of Geol. Soc. India. 36. pp 670-673.

Sahu, K.C. 1990. Heavy metal pollution in mining pp 63-90. Impact of Mining & Environment. Ashish Publishing House. 310 pp.

Saxena, V.K. & S. Vijaya mohan 1993. Concentration of toxic trace elements in deep borehole water at some locations in Hyderabad, A.P. pp 315-320. Inland water resources; India, Ajanta Books International, New Delhi 344 pp.

Shrivastava, J.P. 1986. Radiological Water Pollution. Groundwater News. 5. No.

9 & 10. pp 5-8.

Shvartsev. S.L.; Udodov P.A. and Rasskazov, N.M. 1975. Jr. Geochem. Explor. 4. pp 433-439.

Silvey, W.D. 1967. U.S. Geol. Surv. Wat. Supply Pap. 1535 L. Washington. U.S. Government printing office.

Singh, B.K.; O.P. Pal and D.S. Pandey 1991. Groundwater Pollution : A case study around North Eastern Railway City Station, Lucknow, U.P. Bhu. Jal News 6, No. 2, pp 46-49.

Sridhar, M.K.C. and S.C. Pillai, Jan. 1973, Pollution of Groundwater at Bangalore, Proceedings of the symposium on Environmental pollution. Published by NEERI, Nagpur, India pp 59- 68.

Tamta, S.R., S.L. Kapoor and T. Govardhan. 1991, Manurial N(3 Pollution of Groundwater in the Deccan Formation of Godavari Basin-Ka ·taka. A prel. case study. Bhu-Jal News. 6. 3. pp29-33.

Todd, D.K. 1980. Groundwater Hydrology. Sec. Ed. John Wiley & Sons. Inc. 535 pp.

Trivedy, R.K. 1990. Mining and Environment an overview pp 1-18. Impact of mining on Environment. Ashish Publishing House. 310 pp.

U.S. Environmental Protection Agency, 1977. The report to congress : waste disposal practices and their effects on groundwater : Executive summary. U.S. Environmental Protection Agency. 43 pp.

Verma R.D, 1989. Safeguards for ground water pollution in India, International workshop on Appropriate methodologies for development and management of groundwater resources in developing countries. NGRI, Hayderabad Vol. 2. PP 842-851.

Widyanto, L. et. al. 1975. Paper presented at 2 World Congress on Water Resources organised by IWRA. 33pp.